Teaching Science with Everyday Things

Teaching Science with Everyday Things

Victor E. Schmidt
Professor of Science, Emeritus
State University of New York

Verne N. Rockcastle
Professor of Science and
Environmental Education, Emeritus
Cornell University

AIMS Education Foundation
Fresno, California

This book was set in Times Roman by Waldman Graphics, Inc.
The editors were Phillip A. Butcher and James R. Belser.
The production supervisor was Diane Renda.
The illustrations were done by Raymond F. Houlihan and Mark Walp.
The cover was designed by Margie Anderson.
Creative Graphics was the printer and binder.

TEACHING SCIENCE WITH EVERYDAY THINGS

ISBN 1-881431-57-6

Contents

Preface

This book is designed to be of practical help to teachers, particularly those in elementary schools, and to college students who are preparing to teach. It is intended especially for those who may lack confidence in teaching science or whose background in science may be limited.

The book requires no previous training in science, and no special or costly equipment. It does, however, call for a willingness to try new things, to explore and investigate, and to seek answers through observation and experimentation instead of merely accepting what others say. Above all, it assumes on the part of the reader a desire to help children learn by exploring and investigating the fascinating world around them.

Three underlying themes run through the book:

1 Science—both the process of inquiry and the knowledge that results—is an integral part of education.

2 All learning in science is based, fundamentally, upon firsthand experiences with real things.

3 Science experiences often need involve no unusual, elaborate, or expensive apparatus and materials.

Anyone who expects to find in this book a treatment of subatomic particles, chemical bonds, DNA, galaxies, or the earth's interior is bound to be disappointed. The material that is presented deals with things much closer at hand, things that children—and adults—can watch and touch and try for themselves. Such material hardly requires abstract formulas and complex terminology. However, this is not to say that the content of the book is of a level beneath the dignity and intelligence of college students and college graduates. On the contrary, the principles and approaches developed in the book are basic to the study of science, regardless of level.

A book of this size can, at best, merely touch on some aspects of science and on some approaches useful in helping children become interested and involved in science. Nevertheless, it does include enough subject matter and teaching suggestions to help teachers—especially those with little background—to teach science effectively and to grow in skill and confidence.

In writing this book, we have drawn upon a total of more than three-quarters of a century of experience in teaching science and science education. This has involved working closely with children from preschool on, with high school and college students, with teachers in preparation, and with teachers in service. Throughout, we have tried new and original materials and techniques, and these are emphasized in the book. Still, many of our ideas—probably more than we realize—have come from others, and we can only trust that our contributions will constitute a fair exchange.

We should like to acknowledge our indebtedness to numerous persons—mostly our teachers, colleagues, and students. We wish it were possible to thank each one individually. Perhaps our gratitude to them is best expressed by our sincere desire to share the results of their inspiration and encouragement with a larger number of teachers and, through them, with many more children.

Victor E. Schmidt
Verne N. Rockcastle

In Explanation

POINTS OF VIEW

The first of the chapters that follow, "Points of View," may well be the most important part of the book. In it are presented some fundamental philosophical considerations and general suggestions for making science teaching, especially in the elementary grades, effective and pleasant. These "Points of View" should be reread from time to time and reconsidered frequently insofar as they apply to any particular class, school, and community.

COUNTING AND MEASURING

The second chapter deals with one of the most characteristic aspects of science—quantitative investigation. It presents important objectives and activities that relate to establishing and using units (including metric units), constructing and calibrating measuring devices, making valid approximations and comparisons, and surveying for drawing simple maps. The suggestions are applicable to various areas of science dealt with in subsequent chapters.

AREAS OF SCIENCE

The subsequent chapters relate to major areas of science. In each, an introduction is followed by a listing of some important objectives. These, in turn, are followed by suggestions of instructional activities. There is no special significance to the order of the chapters; one may begin with any of them, or anywhere within one. Consequently, the book can be used in connection with any elementary science program or text—or in the absence of either.

IMPORTANT OBJECTIVES

A sampling of important objectives, illustrative of the goals toward which science teaching should be directed, is presented for each area. The objectives listed are attainable, to some degree at least, by means of the activities suggested for that area. It should be emphasized that these objectives are stated for the teacher's guidance only. In no way does "teaching" them directly, or having pupils "mouth" them, constitute good science teaching.

INSTRUCTIONAL ACTIVITIES

Space permits the presentation of only a sampling of activities. However, these are of sufficient number and variety to illustrate sound approaches to teaching science at any grade level. They involve many of those processes and principles on which modern elementary science programs are based. Although they emphasize *doing,* this does not mean that science teaching should consist exclusively of "experiments" and neglect such activities as reading and writing.

GRADE PLACEMENT

Within each area, the activities are suggested in approximate order of increasing difficulty, roughly according to grade level. Specific grade placement is avoided, however, because this might unduly influence the teacher. For one thing, simple activities are often helpful, even in the upper grades, for children with limited experience. In other instances, pupils in the primary grades may be ready to try some things that are relatively difficult.

QUESTIONS FOR PUPILS

Examples of questions that may be asked of pupils are printed in *italics.* These questions are intended to arouse curiosity and to encourage careful investigation, close observation, and clear thinking. Of course, the suggested questions may be modified and augmented as circumstances require. In general, no answers to them are given, because investigation, observation, and thought—not the book or the teacher—should be the source of the answers.

SOURCES OF MATERIALS

Practically all the items needed for the suggested activities are commonplace and easily available. Many can be salvaged from waste; others may be purchased locally. However, substitute items—perhaps suggested by pupils—often serve equally well, or even better. Occasionally, changes in products or packaging may make some things unavailable; but then, if one considers what qualities made the original items suitable, one can probably find replacements.

SOURCES OF ASSISTANCE

Custodians, high school science teachers, older students, and parents often can provide materials and assistance. However, one should be careful that such persons, in trying to be helpful, do not hinder good teaching. At times they may tend to introduce unnecessarily elaborate equipment or too-complex subject matter and terminology. Or—perhaps even worse—they may "give away" answers prematurely and thus discourage thought and investigation.

POSSIBLE DIFFICULTIES

On occasion, one of the suggested activities may not "work." This could mean that something *new* has been discovered. Or it could be that the materials or techniques used are different from those which the authors had in mind. Generally it is wise to try things oneself, first, with no pupils present. Then, in case of difficulty, one should reread the directions carefully, paying special attention to the materials and techniques suggested—and try again!

Teaching Science
with Everyday Things

Points of View

SCIENCE AND CHILDREN

Children live in a world in which science has tremendous importance. During their lifetimes it will affect them more and more. In time, many of them will work at jobs that depend heavily on science. As voters, they will have a voice in making many decisions that involve science—for example, concerning energy sources, pollution control, highway safety, wilderness conservation, and population growth. As taxpayers, they will pay for scientific research and exploration. And, as consumers, they will be bombarded by advertising, much of which purports to be based on science.

Therefore, it is imperative that children, the citizens of the future, become functionally acquainted with science—with the process and spirit of science, as well as with its facts and principles. Fortunately, science has a natural appeal for youngsters. They can relate it to so many things that they encounter—aquariums, flashlights, tools, echoes, and rainbows.

Besides, science is an excellent medium for teaching far more than content. It can help pupils learn to think logically, to organize and analyze ideas. It can provide practice in communication skills and mathematics. In fact, there is no area of the curriculum to which science cannot contribute, whether it be geography, history, language arts, music, or art!

Above all, good science teaching leads to what might be called a "scientific attitude." Those who possess it seek answers through observing, experimenting, and reasoning, rather than blindly accepting the pronouncements of others. They weigh evidence carefully and reach conclusions with caution. While respecting the opinions of others, they expect honesty, accuracy, and objectivity and are wary of hasty judgments and sweeping generalizations. All children should be developing this approach to solving problems, but it cannot be expected to appear automatically with the mere acquisition of information. Continual practice, through guided participation, is needed.

1

TEACHING SCIENCE

What does it mean, to teach science? The words may seem clear enough. It is possible, however, that they do not convey to the reader the full sense of what the authors have in mind.

Science may be defined as a body of organized knowledge concerning living things, stars, rocks, chemicals, energy, and the like. Actually it encompasses much more than this. Besides *content*—the facts, principles, laws, theories, and hypotheses—science also includes *process*—the observation, experimentation, meditation, imagination, prediction, and other means by which the content is arrived at. And further, science is associated with certain characteristic attitudes and appreciations, among them curiosity, objectivity, honesty, and an abiding sense of wonder.

Teaching is commonly thought of, in the main, as passing on knowledge, but it involves more than this, and certainly more than telling and showing. Ideally, it consists of setting up situations in which learning cannot help but take place. And this learning includes not only the acquisition of facts and concepts, but also the development of skills and habits and of attitudes and appreciations. In this broad sense, teachers inevitably become learners, too, at times—and perhaps can best think of themselves as the senior members of learning teams!

Above all, teaching science should not be envisioned as primarily the presentation of momentous discoveries and esoteric concepts—of dinosaurs, atomic particles, the conversion of matter to energy, and such. Generally it can better be accomplished, in its fullest sense, by means of experiences much closer at hand. Pupils—and teachers—can themselves investigate, for example, the change in sound as water fills a glass, or the apparent shift in position of a coin beneath the glass. Good science, both content and process, can readily be taught with everyday things.

EVERYDAY THINGS

Such things as paper clips, marbles, cardboard, scraps of wood, and rubber bands constitute a veritable reservoir of materials for teaching science. They cost little or nothing and yet are often better than expensive equipment for providing worthwhile learning experiences. They encourage youngsters to become involved and to use their ingenuity, rather than merely to read about things, listen, and look on. Already familiar to pupils, commonplace things like these are not likely to block or distract their thinking in the way that strange and complicated apparatus often does.

Many useful odds and ends are available in quantity, enabling all children to participate in investigations, rather than limiting them to watching demonstrations. Such items permit experiments to be repeated and modified easily. Generally obtainable on short notice, they seldom necessitate requisitions, administrative approval, and the delays that too often postpone the arrival of commercial equipment beyond the interest span of children.

With everyday things, pupils need not stop learning science at the close of school, since such items usually are also available around the home. There the children can repeat the experiences they had in school and then continue with their own ideas and with suggestions from books. Parents, too, may become involved, thus increasing their understanding—not only of science, but also of the school's program.

Everyday things can help to provide for individual differences among children—differences in interest, attitude, creativity, dexterity, and academic aptitude. They can be as intriguing and challenging to slow learners and poor readers as to gifted pupils. They enable youngsters to learn with their hands as well as with their minds; yet they place no ceiling on ingenuity or industry.

GROWTH OF A CONCEPT

A child observes and experiences things and incorporates some of what he or she observes and experiences into a complex mental structure referred to as a *cognitive structure*. This structure is composed of many subordinate, but interrelated and interconnected units or *concepts*. The concepts within the structure usually begin as simple mental images, but in time they become elaborate and inclusive. The concept "ball" is an example.

Suppose that a baby learns that "ball" is something round, red, and bouncy. Weeks or months later, "ball" is also yellow, bouncy, and spherical. (The spherical character may have been included from handling the ball.) A few years later, "ball" may include such characteristics as multicolored, plastic, and common at beaches. And still later, "ball" may include a hard, not-very-bouncy object batted about by a group of older children or adults in a game. Each of these observations may, in turn, be incorporated into the child's constantly expanding, ever more inclusive concept of what "ball" is.

This ever-changing, ever more sophisticated, ever more inclusive concept, "ball," may take years to develop. Its development requires many experiences and countless observations. Depending upon the child, and upon the observations and experiences the child takes part in, "ball" may

remain fairly simple, or it may include Christmas tree ornament, globe, surface/volume relationship, toy, and curled-up kitten. Still, it is a relatively simple concept—not one ordinarily left to the school to develop.

Some children may grasp a concept in a fairly inclusive, if tenuous, form at the outset. Others will develop a concept in a step-by-step, sort of linear fashion. Each child is an individual in that no two will have the same set of experiences or share the same set of observations en route to a particular concept. The teacher should recognize that a concept, however simple it may seem, can be a relatively complex structure, with many possible interpretations by the child.

When a teacher suggests to a learner, "It is like a ball," the teacher must remember that a concept as simple as "ball" already has countless faces. Saying "It is like a ball" to children and hoping that this clarifies something that the children do not understand may actually not clarify at all. It may serve only to suggest spherical, colored, bouncy, or toy, when in fact the teacher may have in mind one end of the upper leg bone, or femur.

Thinking of concepts, and how concepts develop in children, can be an aid to a teacher. However, it is an aid only if the teacher appreciates how tortuous the growth of a concept can be, and how different the same concept can be in the minds of different children.

SEQUENCING CONCEPTS

Seemingly simple concepts such as "ball," complex though they really may be, usually are beyond a teacher's responsibility for teaching. They are preschool concepts. Other, more complex concepts or conceptual structures, such as "life cycle," may be part of a required program and thus part of a teaching responsibility. How efficiently a conceptual structure is presented and learned often depends upon how its supporting concepts are sequenced.

First, it is helpful to consider what are some supporting concepts that go to make up a conceptual structure. For example, the conceptual structure "life cycle" may depend upon an understanding of the following concepts, for even rough comprehension:

"living"	"organism"
"growth"	"adult"
"egg"	"stage"
"seed"	"cycle"

Each of these is anything but simple. Each suggests many observations and experiences over months or years for a child to understand it. The larger, more inclusive structure, however, will be only a word if it cannot be anchored in a reasonable understanding of these supporting concepts.

In sequencing the supporting concepts, the need for one to help explain another should be considered. The concept "electromagnetism" would mean little or nothing to someone who did not understand the concept "magnet." Similarly, the concepts "living," "growth," "organism," "adult," "reproduction," and so on are imperatives for a child who is en route to learning the larger, more inclusive structure, "life cycle."

In teaching about life cycles, that of the fruit fly, *Drosophila*, often is used. Some children think that the adult *Drosophila* is merely a baby fly and that it will grow larger as it gets older. It comes as a surprise to learn that when a fly emerges from its pupal case, it is as big as it will ever get. Big flies must be different from little flies! The concept "adult" must have something to do with "stage," not size.

It is this constant refinement of concepts by the child that leads to enlarging, correcting, and refining the larger, more inclusive conceptual structure. In sequencing the supporting concepts, the teacher can be of great help to the learner. For example, in teaching "life cycle," the concept "circle" may be important. This needs modification, since to complete a circle is to come back to the *same* place. To complete a *cycle* is to return to a similar, but not identical, *stage*. Thus teaching about "cycle" suggests an anchoring concept, "stage." It is in such fashion that a sequence of concepts is planned.

Not all children learn concepts in the same order. Also, what seems to be an efficient order for teaching may not always be the most efficient order for learning. Working out supporting concepts, however, and a reasonable order for teaching them, helps to develop an instructional strategy, makes a teacher more aware of the needs of children, and helps to ensure that important portions of the larger structure are not overlooked.

OBJECTIVES AND OUTCOMES

To be really effective, science teaching—like all teaching—must have clear objectives. Both content and approach should lead to worthwhile outcomes. Neither should be chosen merely on the basis of tradition or just for the sake of novelty.

The more clearly teachers define their purposes, the better and more satisfying their teaching will be. In fact, once definite objectives have been decided upon, planning how to attain these objectives, and how to evaluate the extent to which they have been attained, is greatly facilitated. For example, if one goal is to have pupils be able to connect simple electric light circuits, the choice of materials and methods becomes obvious. If another goal is to have them appreciate the importance of checking measurements, the approach again becomes clear.

In each chapter of this book are listed some objectives. These are grouped in three categories, the order of which indicates the authors' judgment of relative importance. For example:

Attitudes and Appreciations

There is a beautiful orderliness in the motions and changes in the appearance of the sun, moon, and stars.

No creature is, in itself, harmful or beneficial; it is so only in terms of how it affects human well-being.

Skills and Habits

Making a simple magnetic compass, and using it to determine direction

Working with others in carrying out an investigation and in gathering and analyzing data cooperatively

Facts and Principles

Sounds differ in pitch, and this depends upon the rate of vibration, or frequency.

Although earth features such as mountains are continually being changed, the basic materials are not destroyed but are used over and over again.

PLANNED PROGRAMS

Although firsthand experiences, supplemented by vicarious ones, are basic, for good science teaching these experiences can hardly occur in a haphazard, unrelated fashion. They need to be part of a planned program. For one thing, a teacher should be able to expect that in the previous grades the class has had certain experiences and has attained certain goals, on which to build.

To present such a program is not a purpose of this book. Many good programs are already available—in textbook series, state courses of study, and curriculum guides developed by committees of teachers. Some of these programs are outstanding in that they:

1 Have clear and worthwhile objectives—including those that deal with attitudes, appreciations, and skills, in addition to the usual objectives involving knowledge—together with numerous practical suggestions for helping pupils attain these objectives

2 Are sequential, suggesting simple skills and ideas to be developed in the lower grades in accordance with the ability and interest of the children and gradually introducing more complex processes and materials as the pupils grow in maturity and experience

3 Deal with content that is well suited to children instead of crowding down into the grades material from high school and college courses—including highly abstract concepts that have no place in the elementary school

4 Include a good balance of material drawn from the physical, biological, and earth sciences, with no major gaps or pointless repetition and so coordinated with the entire instructional program that science is not isolated from other subjects

5 Require ample time for science—just as for reading, social studies, and mathematics—to permit a wealth of experiences, coupled with reflection, discussion, and fun

FIRSTHAND EXPERIENCES

Firsthand experiences are, in the final analysis, the basis of all learning. A person's education rests, fundamentally, upon direct contacts with the environment—through seeing, hearing, feeling, smelling, tasting, and other sensory channels.

Unfortunately, all too often this truth seems to be forgotten. Too readily do we substitute books about things for the things themselves, films for field trips, and television programs for active participation. Too quickly do we tell, instead of letting pupils find out on their own. Too often do we have them memorize rather than investigate.

Of course, vicarious experiences—those which one has indirectly through the medium of others—are often highly desirable. In fact, in some cases they are the only feasible means of developing concepts. There is certainly no question of the value of books, pictures, recordings, and other instructional aids, but they can never fully replace firsthand, direct experiences.

For example, can merely reading about air pressure substitute for actually feeling it? (See "Air Push," page 33.) Does watching spots of light in a planetarium truly take the place of seeing real stars in the sky? (See "Star Shifts," page 189.) Is listening to someone tell about the delay of echoes nearly as effective as hearing and measuring this delay? (See "Sound Bounce," page 140.)

Can one convey to a child, by words alone, the fragrance of mint, the flavor of wintergreen, or the sting of nettle? How adequate a concept of snow can a youngster have who has never seen snow, felt snow, or otherwise experienced snow?

The same holds true for much of what is taught in school, be it about soil erosion, pond life, solar energy, water pollution, or the methods used by scientists. Firsthand experiences are essential to make these topics truly meaningful to children. A vast number of such experiences are possible, of which this book suggests only a sampling.

PUPIL PARTICIPATION

Inasmuch as firsthand experiences are fundamental to learning science, it follows that every child has a right to a rich variety of such experiences. Every child, therefore, should participate, and no one ought to be, or needs to be, a mere onlooker.

For example, every pupil can make and calibrate a stick balance to use whenever it is needed. (See "Calibration and Confidence," page 24.) When studying magnetism, all the children can work with their own bobby-pin compasses (page 148) rather than just watch the teacher demonstrate. Everyone can have his or her own animals and plants for close scrutiny (as in "Fish-less Aquariums," page 50). The entire class can contribute to the finding of pupils' reaction time. (See "Squeeze Play," page 20.) All can take turns in the rounding of brick pebbles (page 84).

Demonstrations should be used sparingly—chiefly to show pupils what to do or how to do it. However, they are also desirable for those few activities which are too difficult, dangerous, or expensive for everyone to try.

Having every pupil participate may, at times, raise problems, such as those of time, space, materials, storage, and behavior. These problems, however, have solutions that are often obvious if the basic purposes of the teaching are kept clearly in mind. For example:

1 Attempt to teach less, but better, and draw from science content when teaching reading, writing, mathematics, and other skills.

2 Use the playground, lawn, gymnasium, and corridor for science activities, as well as windowsills and chalk trays.

3 Make use of everyday things such as cans, pebbles, nails, and wire—collected with pupils' help, at little or no cost.

4 Keep supplies in boxes, cans, and jars, perhaps on homemade shelves, and then teach pupils to get the items as needed and to put them back.

5 Challenge pupils with interesting and worthwhile activities and there probably will be little mischief—although, perhaps, a bit more noise.

DISCOVERY THROUGH EXPERIMENT

The real essence of science is honest inquiry. Scientists discover new things because they inquire constantly into the unknown. Children, too, can discover by inquiring and experimenting. The things they find out may not be new or startling to you, but to them even small discoveries are exciting.

Many so-called experiments are not experiments at all. Instead, they are verifications of something both teachers and pupils already know. Even so, they may be advantageous for learning, just as repeated trials are desirable. Discovery through experiment is one of the best means of developing:

1 The confidence that comes by finding out for oneself

2 A willingness to try new things, even though the procedures are unfamiliar and the outcome uncertain

3 An open-mindedness that ensures acceptance of a new idea if it proves to be more valid than a former one

The most effective experiments often are those which arise from pupils' own questions instead of from a textbook or a teacher. They motivate pupils much more than artificially imposed experiments and lead to purposeful and hence more efficient pupil activity.

Children's discoveries do not always agree with what others have found. Nevertheless, their discoveries are real and should be considered true until a situation arises or can be arranged where repetition fails to verify their original findings.

When this happens, pupils' changes of mind should come from their own observations, not from adult authority. Scientists do not have a higher authority whom they can ask if their experiments "worked." Like scientists, pupils should learn to rely on what they discover through experiments, not primarily on what the teacher or book says. Their final authority should be the answer to "What does Nature have to say?"

LEISURELY LEARNING

Learning in science should not be hurried. For children to try things, make observations, weigh evidence, come up with alternative explanations, and think of additional examples takes time. If they are to mull over ideas, discuss them with classmates, and test them in new situations, they cannot be rushed.

There seems to be an increasing tendency to cover a large and set amount of material in the grades—a result, undoubtedly, of our concern with measuring pupil performance. This leads to emphasis on the skills of reading and mathematics and on the recall of facts—all relatively easy to test. However, it often results in reducing the time devoted to science—especially to the *process* of science. And the development of this, in particular, is too important to be hurried.

Pupils need time to have firsthand experiences and to think about what they themselves do, instead of merely reading about things and watching what others do. They must have ample time to practice skills such as designing experiments, measuring and recording data, working in groups, thinking critically, and communicating clearly. They must be allowed time to brainstorm and be creative, to develop worthwhile attitudes and appreciations, and above all to enjoy science.

It may be frustrating to let pupils flounder a bit in their search for knowledge and understanding—especially to teachers who know or think they know the answers. Yet, just as a mystery story is spoiled by giving away the solution, the imperfect attempts of children to find out for themselves may be spoiled by telling the answers. Telling saves time—and unfortunately some teachers are more eager to save time than to open minds. However, much of the enjoyment and value of learning science lies in the search, not in the answers!

MEANS OF MOTIVATING

Perhaps the greatest secret of being a good teacher is the knack of getting children to want to participate in learning. Thus motivated, they learn without being coerced because then learning is fun. They may not even realize that it is taking place.

This is not to say that learning need be, or should be, effortless. However, even hard work is fun if it is satisfying. Think of the physical exertion during a ball game, and the mental effort in solving a puzzle or playing checkers.

Pupils become motivated when they participate in activities that are rewarding to them. The rewards, however, must be obvious and immediate; long-range goals have little meaning for young children. The teacher, nonetheless, must keep the ultimate aims in view, realizing that these can be attained, in part, by having pupils do things that are intriguing to them.

Learning experiences with built-in motivation include:

Engaging in physical activity involving large muscles (See "Broomstick Pulleys," page 122, and "Squeeze Play," page 20.)

Watching things move, especially with rapid or repeated motion (See "Reaction Carts," page 125.)

Seeing, hearing, or otherwise sensing pleasant (or unpleasant) things (See "Pebble Jar," page 82, and "Scrap-wood Music," page 135.)

Exercising control over phenomena—making things happen (See "Diving Dropper," page 59, and "Needle Poles," page 149.)

Being surprised or fooled by unusual or unexpected happenings (See "Air Push," page 33, and "Big Finger," page 164.)

Solving puzzles and problems, both physical and mental ones (See "Night Lights," page 154, and "Super Solution," page 73.)

Making estimates and predictions; then seeing how close these are (See "Shadow Motion," page 188, and "Limb Lift," page 48.)

Competing in contests that offer good chances for success (See "Ice-Melting Contest," page 175, and "Goodness! Book of Records," page 106.)

KEEPING RECORDS

"Lest We Forget" might well be the title of a class notebook in which pupils keep detailed records of experiments and other activities. An honest, complete record of observations can be examined at any time to see what really happened. It is surprising how different from the written record a recollection can be!

Some activities, such as "Layers in Logs" (page 47), do not call for extensive record keeping. Others suggest specific ways of keeping records. Drawing silhouettes of "Brick Pebbles" (page 84) encourages pupils to observe more carefully. In "Squeeze Play" (page 20), a column of numbered trials and times is needed for a graph from which the class can generalize. In "Stone Sizes" (page 86), the record is the collection of the objects themselves. In "Thunderstorm Paths" (page 38), the record is a series of numbered marks on a map. Regardless of the manner in which they are made, records serve a single important purpose—to provide an efficient, orderly reminder of pupils' observations.

Some records, such as those of temperature and moisture, may not seem important at the time but are useful in answering questions that arise later. Questions themselves often are worth recording, because they may suggest a different method of procedure or a test of ideas.

Once records are made, they should be used. Pupils soon lose any sense of purpose in record keeping if their records become busywork and are not evaluated at the end of an activity.

By having pupils keep records, even simple ones, from the first grade on, you will help them to make record keeping a habit and an essential part of almost any science activity. Then when they ask, "How do we know?" or "What is the evidence?" their records will give the answer.

REPEATED TRIALS

Pupils rarely learn all they can from a single science experience. They may observe much of what happens. They may even understand the principle involved. But learning to the point of bringing about a change in behavior may result only after repetition.

Some benefits that can accrue from repeated trials are:

1 Improved learning through reinforcement. In "Shadow Motion" (page 188), pupils observe a shadow moving, but only after repeated experiences will most of them learn to predict within reasonable limits how far a shadow will move in an interval of time.

2 Greater validity of results. A class investigating "Seed Surplus" (page 49) will not be able to make a valid judgment on the basis of one sample; therefore, many samples are taken and their combined results are used in arriving at a conclusion.

3 Greater awareness of details and subtle relationships. In "Taller or Wider?" (page 16), pupils making their first measurement often completely overlook heels and hair, but these errors become obvious as they repeat their measurements using different subjects.

4 More opportunity to question, or to suggest new ideas. In "Masses and Motions" (page 124), substitutions of objects or changes in position of and force on an object may lead to testing and modification of a hypothesis until it is accepted as correct.

5 Increased pleasure from the sheer fun of doing. In "Pie-Pan Generator" (page 159), pupils are not satisfied with just a once-only treatment. They like to see the sparks, hear the snap, and light a tube. Such pleasure is excellent motivation for learning.

"What would happen if . . . ?" or "How would a change in mass affect . . . ?" might be overheard as pupils try an activity. Such questions are invitations to repeated trials. The changing of one condition or the substitution of one material for another in an investigation is good science and should be encouraged.

LABELS THAT LIMIT

Jean Piaget once said, "The principal goal of education is to create [people] who are capable of doing new things, not simply of repeating what other generations have done—[people] who are creative, inventive, and discoverers."* Creativity in children is enhanced by a creative teacher. One way to be creative is not to be limited by labels.

One of the most useful objects in the teaching of elementary school science is the drinking straw. A box of drinking straws is labeled "Drinking Straws." But that label may be limiting, because a drinking straw can also be used as a(n):

roller	unit of measure
sighting tube	tube for stick figure
aquarium siphon	stirring rod
blowpipe	musical pipe

When you need to use materials or objects for investigation or demonstration, think in terms of the *purpose* to be served, and do not be limited by the name or label ordinarily applied to the objects. Suppose, for example, that you need a container to hold water. A glass tumbler or beaker would do, but perhaps a paper cup or even a discarded soft-drink can might do as well in a particular situation.

In undertaking an experiment or investigation in which a conductor is needed, you might think of a copper wire. If you do not have a copper wire, consider using a strip of aluminum foil. Or a paper-clip chain. Or a piece of wire screen. Or even an uncoated metal zipper that has been zipped together!

If you need insulated wire, remember that a drinking straw is hollow and is a nonconductor. A rolled-up strip of foil inside a drinking straw is an insulated wire—if you are not limited by labels.

Sometimes the fun of investigation lies in making do with ordinary materials in extraordinary ways. The teacher—and the child—who can find alternative uses for everyday things is likely to find solutions to problems that would elude a person who is limited by labels.

*Richard E. Ripple, and Verne N. Rockcastle, *Piaget Rediscovered*, Cornell University Press, Ithaca, N.Y., 1964.

Counting and Measuring

Among the most important process skills to be developed in elementary school are observing and communicating. Children may go through life without having developed certain process skills, but not these. Basic as these are, however, they draw heavily upon other skills, such as counting and measuring. When children observe and communicate with others, they must describe accurately what they have seen. Accurate description often depends upon counting or measuring. When children compare accounts or results of investigations, much rests on counting and measuring.

Human progress has been due, in large part, to the ability to measure with greater and greater precision. A part for a modern machine must be close-fitting, and its replacement must be nearly identical. The amount of medicine in a pill, the curvature of a lens, and the position of the components in a printed circuit—all require precise measurements. Children cannot work to this degree of precision in their explorations, but they can appreciate the need for precision and will refine their own crude instruments and techniques, even in simple activities.

The activities suggested here should not be considered ends in themselves. Instead, they are a means of introducing children to ways of measuring—ways that have meaning in a child's world. The devices described are not gadgets; they are instruments based on sound scientific principles, but at a level children can understand and use in investigating their environment.

Just as learning to read a thermometer is a step toward understanding weather, making an unequal-arm balance (see "Calibration and Confidence," page 24) is a step toward understanding how mass can be measured. Finding the mass of a slice of apple should never become the primary objective of the teacher, even if it is the primary objective of the pupil. For the teacher, it should merely illustrate a method of obtaining information and should introduce the concept of relative mass. The teacher's long-range objective should always be more inclusive than that of the child.

SOME IMPORTANT OBJECTIVES

Attitudes and Appreciations to Be Encouraged

Events or objects are best described if they can be counted or measured accurately.

Measurements are never exact, but some are more nearly exact than others.

Some measurements must be made with more precision than others, because of their purpose.

When measuring, one should be honest and careful and as precise as the instruments being used will permit.

Indirect measurements can be as accurate and useful as direct measurements.

No one system of units is inherently more accurate than another, but the metric system is used by most people and is the most convenient.

Good measuring instruments are valuable, and they deserve the care one provides for valuable things.

A measurement checked several times, preferably by several observers, is better than a measurement taken only once by one observer.

A single system of units is desirable because it makes international communication easier and more meaningful.

Good measurements help one develop a healthy confidence in one's findings.

Skills and Habits to Be Developed

Holding, and reading to the smallest division, simple devices to measure distance (rulers, meter sticks, folding rules, etc.)

Reading various scales at right angles so as to avoid parallax

Weighing, or finding the mass of, objects and timing events with accuracy

Changing measurements in one unit (such as meter) to another unit in the same or a different system (such as centimeter or inch)

Finding the arithmetic average of several measurements or counts

Estimating, with a fair degree of accuracy, various distances, masses, and times

Visualizing units, such as square centimeters in a square meter and milliliters in a liter

Finding the dimensions of very small objects or time intervals either by using indirect means or by finding the total dimension of many and dividing by the number involved

Facts and Principles to Be Taught

Dimensions of ordinary measurement are distance, mass, and time. Most common measurements are one or a combination of these.

Measurements are really comparisons with other known or accepted dimensions.

In order for people to communicate measurements to one another and to understand what they mean, there must be standards for comparison. International standards are carefully made and protected.

Mass is a measure of the amount of matter in something. It is not dependent upon the pull of gravity.

Weight is a measure of the attraction between an object's mass and the earth. If two objects on scales are "balanced," the attraction between the earth and each of the objects is the same, so the objects are of equal mass.

The weight of an object can also be found by measuring how much it bends or stretches something springy.

The number of swings a pendulum makes in a certain time depends mostly on its length, not on its mass or the distance through which it swings.

STANDARDS OF MEASUREMENT

To show pupils that there is nothing sacred about systems of measurement, let them develop a system of their own and use it. This will help them to appreciate how our own conventional units were developed, what their limitations are, and how the systems we use depend on a few standards chosen more or less arbitrarily.

First, let the pupils select some object as a *standard* unit of length. It may be a drinking straw, an unsharpened pencil, or a piece of paper. Let them give it a name, too, such as "gort." Then let each pupil make a copy of the standard gort from a strip of paper or plastic.

For a unit of mass, the pupils should decide on another standard. It may be a pebble, a nail, a particular block of wood, or a gort of clothesline. It should have a name, too, such as "lob." Once the lob has been established, it can be set on a balance, and another lob can be made by filling a plastic bottle with sand until it balances the stan-dard. In a similar manner, 2-lob, 5-lob, and 10-lob masses can be made. *How can a ½-lob or a ¹/₁₀-lob mass be made?*

A "blip" may be the time required for a pendulum 1 gort long to make one swing to and fro. (See "Swinging Second-Timer," page 19.) For a longer time interval, a small hole can be punched in the bottom of a container. Then the container can be filled with water to a level that will take, say, 100 or 1,000 blips to empty. *Will it really empty?* (See "Empty or Full," page 31.)

When the system of units is established, it will become apparent that it is based on a few objects chosen arbitrarily. If these were lost, it would be difficult to reconstruct the system. For this reason, the class should consider keeping their standards in the school safe (their Bureau of Standards). It will also be apparent that descriptions of objects in terms of the class standards are possible only if other people know what gorts, lobs, and blips are. That is why a single system for all people is desirable.

UNITS FOR USE

Sometimes units of measurement that pupils read about—metric as well as older units—do not mean much to them. One way to help correct this is to let them make a labeled collection for ready use and reference—like utensils in a well-ordered kitchen.

They may start with just a few units and measuring devices, and then add more as the need arises. Some that may be included are:

1 Homemade metric rulers. (Long strips of cardboard or heavy paper that pupils have marked off in centimeters and millimeters.)

How tall are you, in centimeters? How wide is camera film, in millimeters?

2 Meter-long sticks. (Pieces of scrap wood sawed to a length of 100 centimeters, or 39⅜ inches, or yardsticks extended to this length by taping on strips of wood.)

How far is 100 meters? In what time can you run this distance?

3 Lengths of 10 feet and 10 meters. (Heavy string, or inexpensive chain or rope from a hardware store, cut while pulled taut. A chain like this, however, must not be confused with the surveyors' unit called a *chain,* which is 66 feet long.)

About how many acres, or hectares, in size are the school grounds? (A square-shaped acre is very nearly 209 feet on a side. A hectare is equal to 10,000 square meters.)

4 A marked-off distance of 1 kilometer. (Two painted lines or other markers 1,000 meters apart, measured with a bicycle. See "Wheel Measure," page 18.)

In what time can you walk this distance? How long is your average step?

5 Flat surfaces with areas of 1 square yard and 1 square meter. (Square pieces of cardboard, wallboard, or carpeting—or squares marked off on the floor—with sides 1 yard, and 1 meter, long. Also, a few rectangular pieces of odd shapes, with areas of 1 square yard or 1 square meter.)

How many square centimeters are there in 1 square meter? Have pupils mark off some.

6 Frames that outline volumes of 1 cubic yard and 1 cubic meter. (Twelve yardsticks and twelve meter-long sticks with their ends glued to eight corners of cardboard cut from cartons.)

About how many of you could fit in a cubic yard of space? Into how many cubic kilometers might everyone on earth fit?

7 Blocks whose volume is 1 cubic centimeter. (Cubes of potato, soap, or soft wood carefully cut so that each edge is 1 centimeter long.)

Is a sugar cube more or less than 1 cubic centimeter in volume? About what volume of sugar do you eat in a day?

8 Containers with capacities of 1 liter and ½ liter. (A "quart" and a "pint" bottle or jar that actually hold 34 and 17 fluid ounces. See "Liters, Quarts, and Quasi-Quarts," page 16.)

In most countries, milk and gasoline are sold by the liter. *About how many liters of each of these does your family use in a week?*

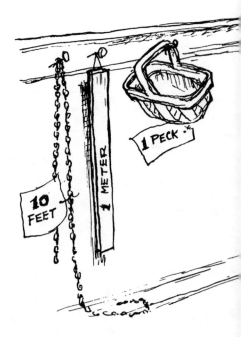

9 Liquid containers calibrated in fluid ounces and milliliters. (Measuring cups, calibrated bottles, and medicine cups.)

Quick, how many fluid ounces are there in 3 quarts? How many milliliters are there in 3 liters? Is it easier to convert from quarts to fluid ounces, or from liters to milliliters?

10 Dry-measure containers of various sizes. (Bushel, peck, quart, and pint baskets and cartons.)

About how many apples or potatoes are there in 1 bushel? In 1 peck? Have pupils estimate the number first, then count.

11 Weights of 1, 2, 5, and 10 pounds. (Plastic jugs filled with sand to make them weigh these amounts, then sealed.)

How well do household scales agree? Let pupils test some with these "standard" weights. *What would be good "standard" weights for testing scales used to weigh people?*

12 Masses of 1 kilogram and ½ kilogram. (A stone or a piece of brick that weighs as much as 1 liter of water, and something half this heavy.)

Which would last longer in your home, 1 kilogram of butter, or 2 pounds? What is your *mass, in kilograms?*

13 A set of masses, in grams. (Coat-hanger wire cut to weigh the same as two new U.S. nickels—10 grams; and an identical piece marked off in 10 equal lengths and cut at the first, third, and fifth marks. This gives masses of 1, 2, 2, 5, and 10 grams. The wires may be bent to show their mass—the 2-gram wire, for example, in half; the 5-gram wire like a W with one additional bend.)

What is the mass of a marble, in grams? Of a penny? See "Three-Corner Balance," page 27.

TALLER OR WIDER?

With their arms stretched out at the side, are people taller than wide, or wider than tall? It is an interesting experience for your pupils to write what they think and then test their opinions by measuring everyone in the class.

How many think that tall, slender people are taller than wide? Wider than tall? Equal? What about short, stout people?

Divide the class into teams, each with a reporter who will write down questions and comments made by the pupils as they investigate. The reporter can also take down the measurements. When all the measurements have been made, let the class listen to the questions and comments made by members of each team as they went about the investigation. *What questions and comments suggest that you became more precise in your observing and measuring as you went along?*

Ask each team to describe how it ensured that measurements were uniformly made, and whether the team made any changes along the way to minimize errors. *Were any measurements made before asking that shoes be removed? How was the top of each person's head projected against the wall so height could be measured? If a book was used, how was it held to make sure it was kept level?*

Ask the class if it makes a difference whether the subject is standing or lying down. *Will a person's measurements be the same in the morning as in late afternoon? What difference does age or sex make?*

Pupils often are quick to generalize on the basis of a small sample. ("*Everyone* does it!" or "*All* the kids have them!") Measuring whether people are taller than wide or wider than tall will almost surely make pupils more critical of first guesses. It will make them suspicious of data not carefully collected or of data collected for one group and then generalized to include other groups. It will also make them reluctant to generalize without plenty of evidence.

LITERS, QUARTS, AND QUASI-QUARTS

Throughout the world, most people use the metric unit *liter* as a measure of volume, but in the United States many people still cling to quarts. *How does a* liter *compare with a* quart? *Further, are what we call "quarts"—of milk, of ice cream, of strawberries—really* quarts? Let pupils find out.

First, show the class how to use a measuring cup properly. Fill it nearly up to a mark with water, and set it down so that it is level. Then hold your eyes level with the mark, and slowly add water until the *flat surface* of the water is even with the mark.

Now have pupils test themselves. Let each, in turn, carefully measure 1 quart of water, pour it through a funnel into a "quart" bottle with a long neck, and mark the level on a strip of adhesive tape. *How well do the measurements agree? Can you read a measuring cup as accurately when you hold it up as when you rest it on a level surface?*

Then let teams test different measuring cups. *Do these cups agree? Do 4 cups, or 2 pints, measured separately, add up to exactly 1 quart?*

Pupils may also check bottles, jars, jugs, and other containers. *Do these hold the exact volumes marked on them? Can you find any that are marked in metric units?*

Teams may also investigate these questions:

● *Do "quart" ice cream containers have the same volume as "quart" milk cartons? How much do "quart" berry baskets really hold?*
● *How does the dictionary define* quart? *Is a dry quart the same as a liquid quart?*
● *If a* liter *of milk, instead of a* quart, *were shared equally by four of you, would you notice the difference?*
● *What is the mass of a liter of water, in grams? How much does a pint of water weigh? Just what is a fluid ounce?*
● *How could you test the accuracy of measuring cups at home? How might they be designed so as to be more accurate, yet still convenient to use?*

AN ENVIRONMENTAL SAMPLE

A class often keeps an aquarium or a terrarium to represent some part of the environment—pond, lake, woods, or desert. In an aquarium may be kept some fish, a few tadpoles, a turtle, or some aquatic insects. A desert terrarium may have one or two lizards and some cacti. A woodland terrarium may have mosses, some small seed plants, a toad or frog, and one or two salamanders. The children think of these aquariums and terrariums as representative samples of the habitats that they are supposed to illustrate. But *are* they? A simple activity will show that they probably are not, and will point up what really would be representative.

First, have the class record the kinds and numbers of organisms in each aquarium or terrarium. The tally might look like this:

Aquarium	Terrarium
3 fish	2 lizards
2 tadpoles	1 snake
1 turtle	4 cactus plants
5 aquatic plants	

Next, have the class cut a piece of newspaper to equal the area of the aquarium or terrarium. Then, using either similar pieces of paper or the measurements of the one piece, find out how many such pieces would fit on the floor of the classroom.

Suppose that an aquarium or terrarium is 0.25 meter wide and 0.50 meter long. That would give it an area of about 0.12 square meter (0.25 × 0.50). A typical classroom is about 8 meters wide and 12 meters long. It has an area of about 96 square meters. The typical aquarium or terrarium would fit into this classroom about 800 times.

Now let the pupils determine how many fish, turtles, lizards, cacti, or other organisms that they have in their aquarium or terrarium would fit into the classroom if the aquarium or terrarium was a truly representative sample. For example:

If the aquarium or terrarium has:	In the area of the classroom would be:
3 fish	2,400 fish!
2 lizards	1,600 lizards!
1 turtle	800 turtles!
4 cactus plants	3,200 cactus plants!

Could an area the size of the classroom ever have this many organisms? Is the classroom aquarium or terrarium really representative of the environment that it is supposed to represent?

Let the class make up an aquarium or a terrarium that is representative of a pond or a desert. When the pupils have decided what to put in it, do not be surprised if there are only a few insects and no fish in the aquarium, or no lizards and perhaps only one cactus plant in the terrarium!

WHEEL MEASURE

Pupils have many opportunities to measure and to become adept at estimating short distances. But longer ones—such as the distance around the school, parking lot, or block—not only are difficult to estimate but take a long time to measure with a ruler or tape. However, pupils can have fun measuring long distances quickly and fairly accurately by means of wheels. Let them write down how far they think it is from one end of the school building to the other and then check it as follows.

Ask that a few bicycles be brought to school. Divide the class into groups, each with a bicycle, a metric tape measure or a string and meter stick, and a piece of chalk. Take them to a sidewalk or parking lot and there let each group measure some distances as follows.

Stand the bicycle upright and let someone make a mark on the front tire where it touches the pavement. Make another mark directly next to it on the pavement. Then push the bicycle in a straight line

until the chalk mark on the tire meets the pavement again, and mark this place. *How far apart are the two chalk marks on the pavement, in meters? How does this distance compare with that around the outside of the wheel?* Measure to be sure. *From your measurements, how far would the bicycle wheel travel in 10 turns? In 100 turns?*

It is easier for a bicycle rider to count the number of turns a pedal makes than to count the turns a wheel makes. So let each group repeat the above activity, but this time find out how far the bicycle travels for each complete turn of a pedal. (For 5- and 10-speed bicycles, there should be a measurement for each of the normal touring speeds.)

As an after-school activity, each pupil can record on a piece of tape stuck to the handlebars how far his or her bicycle travels for each turn of the pedal. (For multispeed bicycles, the tape should show more than one gear and distance.) Finding distances by bicycle can be useful in many instances. (See, for example, "Sound Bounce," page 140.)

SWINGING SECOND-TIMER

A most useful tool in science is a device for measuring time. One of the best is a pendulum—especially one that swings exactly once each second. *Every* pupil should have the fun of making and using one of his or her own, as follows:

1 Tie one end of a thin thread to a metal washer or nut.

2 Knot a loop in the other end of the thread.

3 Stick a pin partway in between the pages of a book that extends beyond the edge of a desk.

4 Hang the loop on the pin so that the washer or nut can swing without interference.

5 Finally, pull the pendulum to one side and let it go.

How many complete, back-and-forth swings does the pendulum make in 1 minute? A Timekeeper for the class may say "Ready! Set! Go!"—and then call "Stop!" after exactly 1 minute. Each pupil should record the number of swings, and check it. *Is it the same each time?*

Perhaps only one or two of the pendulums made by a class will happen to swing exactly 60 times per minute. But the others can be changed until they, too, swing at this rate. *Does making a pendulum shorter cause it to swing faster or slower? Does making it swing harder and farther affect its rate?*

A pendulum like this is disturbed by wind. Therefore, for outdoor use it may be hung inside a tall glass jug or bottle. The thread may be put through a nail hole in the cap, and taped on top.

With second-timers like these, pupils can answer such questions as:

1 *In what time can you run a 100-yard, or a 100-meter, dash?*

2 *How fast can toy cars go, in feet or in meters per second?*

3 *Do drivers passing the school obey the speed signs?* (At 15 miles per hour it takes a car 5 seconds to go 110 feet.)

4 *What is the speed of sound in air?* (See "Sound Bounce," page 140.)

5 *How quickly can you get a marble or a nut through a maze?* (See "Mass in a Maze," page 129.)

6 *How fast does sand settle in water?* (See "Sorted Stones," page 87.)

7 *How long does a candle burn when covered by a jar?* (See "Flame Life," page 71.)

LOOP KNOT

SQUEEZE PLAY

Scientists often need to measure things so tiny that ordinary measuring instruments are not of much use until many of the tiny things are added together. Finding the thickness of a sheet of paper is one illustration of this; a ruler can be used to measure the thickness of 100 sheets, but not 1. Similarly, a watch can be used to measure the total reaction time of a class, but not of one pupil.

Ask the class to form a line around the room and hold hands. Stand at one end of the line and clasp the hand of the child next to you. In your other hand hold a watch with a sweep second hand. Explain that you are going to squeeze the hand of the child next to you when the second hand gets to the top of the dial. As soon as the child feels your squeeze, he or she should squeeze the hand of the next child, and so on to the end of the line. The last child to feel the squeeze should call out so that you can note the time that has elapsed.

Do this several times, each time plotting the result as a graph on the board. The second trial probably will take considerably less time than the first. There will be less difference in the third trial, and still less in the fourth. The graph will level off at the minimum time for the whole class's reaction. If it takes, say, 10 seconds for the reaction time, and there are 20 pupils in the line, then the average reaction time is ½ second per pupil (10 seconds divided by 20 pupils). *What would be the average reaction time if 7 seconds were required for the squeeze to be passed along by 28 children?*

How does the average reaction time for a left-to-right squeeze compare with that for a right-to-left squeeze? If the pupils decide to try it, remind them that they have already *practiced* in one direction and that might prejudice the results. Let them discuss a fair way to find out which direction is faster. *How might a test be set up to determine whether reaction time is faster when the class is rested than when the class is tired?* Suppose the class makes a graph of many trials, using the same conditions for each trial. *How could such a graph show when the class is getting tired?*

Reaction time measured in this way is short. That is because the pupils are expecting the squeeze. But when someone is not expecting a stimulus, his or her response is not nearly so quick. A car driver, for example, usually does not react for about ½ second after something happens on the road ahead. At a car speed of 90 meters per second, the car would move about 45 meters before the driver could take his or her foot from the accelerator. Another 45 meters or so before the driver could touch the brake pedal. And several more before the car could be braked to a stop!

At the average reaction time measured for the class on the first trial, let the pupils measure on the parking lot how far a car might move if that same time was needed for:

1 driver to react, plus . . .
2 foot to move to brake pedal, plus . . .
3 car to be brought to a stop. (Three times the average reaction time, in all!)

TRIPLE-CAN BALANCE

For thousands of years people have compared the weights of objects by using balances. Today, these are still valuable tools of science.

A very simple balance can be used by youngsters for rough weighing. To make one, get three clean cans—a large, undented juice can that has been opened by punctures, and two shallow cans of equal weight. Tape a large nail to the large can, opposite its seam. Then make two nail holes in each shallow can, put the ends of two equal lengths of string through them, and tie knots.

Now stand two smooth wood blocks of equal height on edge. Set the large can on them so that it can roll freely, its seam coming to the top. Finally, drape the strings over the large can, and tape the center of each string to the seam.

Which of two objects, such as oranges, is heavier? A child need only put one in each hanging can.

How much does something weigh? The child may put it in one can and add weights—possibly identical nails—to the other can until the two balance each other. Then the object's weight can be stated in "nail-weight" units.

BOUNCE HEIGHT

Fractions are an important mathematical expression, usually introduced in intermediate grades. Often the use of fractions is limited to the mathematics period, with little application to problems of direct interest to children. Decimals, like fractions, often are limited to mathematics exercises. Both fractions and decimals can find a use—an interesting and challenging use—in determining the bounce height of a ball, as follows.

Divide the class into small groups, each with a meter stick and a bouncy ball such as a ping-pong ball or golf ball. Let each group drop its ball from a height of 1 meter, measuring the height to which it bounces on the first bounce. Have them do this several times and find the average or *mean* height of the first bounce. Questions that may arise, and that the children themselves can answer, are these:

Should the ball be held so the upper, or the lower, surface is 1 meter above the floor?
When we measure how high it bounces, do we measure to the top of the ball, or to the bottom?
How can we tell when the ball has reached the highest point of its bounce?

When the groups have determined how high the ball bounced on its first bounce, let them write this as a fraction or a decimal of its drop height. If it should bounce 75 centimeters, then its bounce height would be ¾, or 0.75.

Next, ask each group to find the fraction or decimal for the second bounce—compared to the first bounce height. Suppose, for example, that the ball bounced 56 centimeters on its second bounce. Then it would have bounced $^{56}/_{75}$ of its first bounce height, or ¾ or 0.75 again. It may be that the ball seems to bounce the same fraction or decimal each time. Or it may be that the fraction or decimal changes. Ask each group to keep a record for several bounces, each time finding the fraction or decimal of the preceding bounce height.

Then, with the whole class listening by cupping their hands behind their ears (see "Louder Sounds," page 136), drop a ball from a height of 1 meter and see how many bounces it makes. *From the pattern of measurements for that type of ball, how high would it have bounced on, say, its tenth bounce? Its twentieth bounce? Its last bounce?*

An interesting associated problem: When the class counts the number of bounces, they will count to themselves, "One . . . . two . . . . three . . . four . . five . sixseveneightnineten . . ." and will lose count because *words* are used to register bounces. Counting is always done by words. When events happen too fast for the words to be pronounced, even silently to one's self, then the count is lost. Let the class discuss, think about, and try various ways of counting rapidly occurring events. Someone in your class may find a new and different way! There *are* things to be discovered, even in the most ordinary and accepted ways of doing things. One of the most enjoyable and challenging results of investigating interactions in your classroom, and in science as scientists do it, is finding new ways of doing things. Creative ways! Ways other people never thought about!

CALIBRATION AND CONFIDENCE

Children may think that the marks on a thermometer, or on a balance, are put there by someone who just *knows* where they should go. But almost every scientific measuring instrument has been *calibrated*, which means putting marks in the right places for that particular instrument. Instruments that have not been calibrated may be unreliable. Calibration helps to ensure accuracy.

Let your class calibrate some unequal arm balances, using these directions for each one. First, get a slender, flat, uniform stick about 40 to 50 centimeters long. At one end tape a hook made from a bent paper clip. From this hook, hang a large paper cup so it is free to swing. Then balance this stick-and-cup on some cylindrical object such as a small juice can. At the balance point, make a mark across the stick and label it "0," meaning that there are no masses in the cup.

Next, select some standard masses such as identical nails, washers, or nickels to calibrate the balance. (A nickel has a mass of 5 grams.) Add one unit of mass to the cup, balance, mark the balance point, and label it "1." Repeat this procedure, adding one more mass each time to the cup. When the balance has been calibrated, it can be used to find the mass of other objects.

Are the calibration marks equally spaced? Suppose that you lost the cup and had to use a different kind on the balance. *Could you still use the same calibration marks? Why do you think so? Would the spacing of the marks on one stick be the same for any other stick?*

Some things to do and to find out with a calibrated balance:

1 Peel a banana. Find the mass of the peel, and then of the edible part. *Compared to the total mass of the banana, what fraction or percentage is edible? Compared to the cost of the whole banana, how many cents' worth of edible banana is there?*

2 It is said that animals and other living things are mostly water. After a rainstorm, when there are drowned earthworms on the sidewalk, collect some and dry off the rain water. Put them in the cup and find their recently living mass. Then lay them on paper towels to dry for several days. When they are brittle-dry, find their mass again. *From the two measurements, how much of the earthworms was water?*

3 Some people peel potatoes before cooking them. Not only is much of the food value of a potato discarded in the peels, but some of the mass of the potato is discarded as well. *How much of a potato is lost by peeling?*

4 Sometimes it is important to find the area of a city, park, or farm. When the boundary is irregular, area is difficult to measure—but it can be found by using a balance. Prick the outline of a map onto a sheet of cardboard. Cut out the outlined shape. Find its mass. Then use the scale of distance on the map to outline one square unit of area on the same kind of cardboard. Cut this out and find its mass. *From the* ratio *of the two masses, what is the area shown on the map?*

FASTER, FASTER!

Today's pupils observe, directly or indirectly, many objects that *accelerate*—space vehicles lifting off, dragsters zooming from a starting line, bicycles getting underway from a position of rest, and runners leaving their starting blocks at the sound of the gun. Expressions such as "gunning it" are associated with acceleration—going faster and faster. Yet most upper-grade students still have a fuzzy notion of what acceleration really is. Letting them *feel* their own bodily acceleration in a measured, timed way, as follows, may help to clarify their concept of acceleration.

Take the class outdoors to a sidewalk or parking lot. There, have the pupils make chalk lines or tape lines across the walk to mark successively greater intervals. Beginning with a "start" line, the marked intervals should be ⅛ meter, ½ meter, 1⅛ meters, 2 meters, 3⅛ meters, and so on. Each interval should be ⅜ meter longer than the preceding one. A total of 10 intervals will be more than enough.

Next, set up a 1-second pendulum that a pupil acting as a Timekeeper can observe. (See "Swinging Second-Timer," page 19.) Ask that pupil to clap his or her hands loudly at 1-second intervals, according to the pendulum's swing.

Now have the pupils, a few at a time, stand at the start line, ready to move forward at the sound of a clap. At the next clap (1 second later) they should be crossing the next line (⅛ meter away). At the next clap (1 second later), they should be crossing the next line. Another second, another line, and so on. Because each successive interval of distance is ⅜ meter longer than the preceding one, the pupils will be accelerating ¼ meter per second each second. When they can no longer accelerate at this rate, they will fall behind in matching claps to line crossings.

What is the longest time that any one of the pupils can accelerate at ¼ meter per second each second? Then how fast would that pupil be going, in meters per second? Suppose lines were made at 0, ½, 2, 4½, 8, and 12½ meters (representing an acceleration of 1 meter per second each second). *What is the greatest number of lines the fastest pupil can cross before falling behind in matching claps and line crossings?*

To show the acceleration of gravity, let the class draw a new "G" line 4.9 meters from the start line, and a second G line 14.7 meters beyond the first one. To accelerate at the rate gravity pulls a falling object, one would have to cross the first G line in 1 second, and the next G line in 2 seconds! An acceleration equal to "G" is very fast indeed!

SURVEYING WITH STRINGS

Having a class make a simple map of a small area by surveying it with strings helps greatly in teaching what maps are and how to read them. It relates science, mathematics, geography, and art!

Select a flat area with trees and posts or other fixed objects on it, perhaps part of the school grounds or a park. Set a small but sturdy table in the middle of this area, and tape a large sheet of cardboard on top. Then draw a tiny rectangle at the center, to represent the table itself on the map. For a large class, set up a few tables like this, some distance apart. **Ask the pupils not to lean on them, and not to move them, even slightly.**

Now let small teams of pupils survey the area, using strings exactly 10 feet long. Start by having two teammates hold a string taut, directly across the tiny rectangle and in the exact direction of a tree or other fixed object. Then let a third teammate check by sighting along the taut string. *Is it over the rectangle, and exactly in line with the tree or other fixed object?*

If so, a fourth teammate should mark (*A*) where the string crosses the edge of the cardboard, and then rule a straight line (*B*) from this mark to the rectangle. This is the *line of sight* to the tree or other fixed object.

How far, in the direction of the line of sight, is the fixed object from the table? The team should measure this distance, using 10-foot strings, and record it in ''string lengths'' or feet (*C*).

Each team, in turn, should then repeat these steps (*A* through *C*) for a few fixed objects, drawing a line of sight to each one and recording its distance. Then all the teams should decide on a *scale* to represent distances on the map—for example, one space on lined paper may be used to represent 1 string length or 10 feet.

Using this scale, pupils should mark off, along the line of sight, the scale distance (*D*) to each object to be shown on the map. Then they should draw a suitable symbol (*E*) to represent the object. And, finally, they should add the scale (*F*) and directions (*G*) found with a compass. (See ''Bobby-Pin Compasses,'' page 148.)

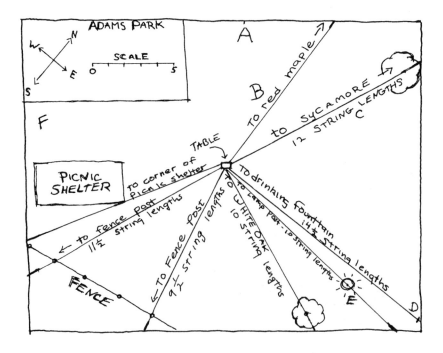

THREE-CORNER BALANCE

A device for rough weighing may be very simple. (See "Triple-Can Balance," page 21.) For many purposes, however, a much more sensitive balance is needed. A pupil can make one as follows:

1 Cut a flat, unbent piece of corrugated cardboard, preferably with a paper cutter, to make a rectangle 15 by 30 centimeters. Have the corrugations run lengthwise.

2 Mark the exact centers of the long sides of this rectangle. Then rule a line between them.

3 Rule lines from the center of one long side to each of the far corners. Now cut along these lines carefully, to make three triangles.

4 Push a needle with strong thread straight through the middle triangle, 1 centimeter from its longest side, on the ruled line. Tie the thread to make a long, *single* loop.

5 Similarly, put threads through the matching corners of this triangle, each 1 centimeter from its longest side and exactly 12 centimeters from the center thread. Tie them to identical paper cups.

6 Hang the loop of thread from a stick resting across two desks. Be sure that the balance hangs completely free.

7 Tie a thread to a metal nut or washer and suspend it from the stick. Have the weight hang just below the bottom corner of the triangle.

8 Check whether the balance hangs level, with the ruled line vertical—parallel to the thread holding the weight. If it does not, snip off bits of cardboard from the heavier side to make it balance.

How sensitive is this balance? What happens when you put a paper clip in one cup? Then how many pins are needed to balance the paper clip, so that the ruled line becomes vertical once more?

A lightweight balance like this has many uses. (See "Dough Raiser," page 78, and "Seed Surplus," page 49.) It may be fitted with paper bags in place of paper cups, to show that gases, too, have weight. (See "Flame Killer," page 72.)

Air and Weather

Some people may think that air is too tenuous for study in the elementary school classroom. Pupils cannot see it, smell it, or taste it, nor can they feel it, or weigh it, as they can a pebble. Yet air is one of the most essential substances in a pupil's environment. In spite of its being invisible, it *can* be investigated, and it is fun to do so.

Learning about air not only helps pupils to appreciate and understand the vital fluid in which they live and the weather that affects much of what they do, but also helps to show them the methods of a scientist. In investigating an invisible substance, children can engage in the kind of organized groping a scientist must do when he or she probes the unknown.

At first, children should simply investigate what air is like and where it is found. To sense that air is real, a beginner must capture it, squeeze it, carry it, and sense its resistance to being pushed aside. A child ordinarily does not think of a room as being full if there is "nothing" in it. It takes time and many experiences with air as sensible stuff before it becomes as real as a pebble.

Next, children should have ample opportunity to investigate movements of air, air temperature and its changes, air pressure, and the role of moisture in the air. As basic understandings develop, pupils can study changes in weather and can begin to hypothesize the causes of these changes and to recognize weather patterns and tendencies.

Although some investigation and prediction of weather is desirable, it should not become a prime objective to make forecasters of pupils. Forecasting weather is a complicated business, one that even professionals find difficult and at which they are not always successful. Still, so much of human activity depends upon weather that pupils should be aware of its changes and the causes of some of them, and may well try their skill at prediction in a modest way.

SOME IMPORTANT OBJECTIVES

Attitudes and Appreciations to Be Encouraged

Air is as important a natural resource as water or soil, and should be valued to the same extent; it should not be polluted or taken for granted.

Air and its moisture are responsible for some of the most awesome and beautiful displays in nature, including thunderstorms, hurricanes, rainbows, clouds, mirages, blizzards, and glaciers.

Activities such as flying kites, sucking milk through a straw, and skiing on snow would be impossible without air.

Weather is the result of naturally occurring events and is not dictated by some supernatural power that operates with caprice.

The social and economic development of an area depends largely upon the weather and climate of that area.

Those of us who rely on special clothing and heated homes cannot fully appreciate the extremes of weather faced by some people and by many plants and animals.

Human understanding of weather and the ability to predict it depend, in part, upon careful observations and accurate records.

The observations and inferences that one can make in studying an invisible substance can lead to as convincing a model as those that one can construct in studying something visible.

Weather forecasters are responsible people who try to understand complicated, incompletely understood processes. As they learn more about causes of weather, their predictions will improve.

Skills and Habits to Be Developed

Reading accurately, to within one unit, the scales of thermometers and barometers

Recognizing air as a fluid that must be displaced when filling, or that must be admitted when emptying, a container of liquid

Recognizing cumulus and stratus clouds and the weather usually associated with each

Constructing simple weather instruments such as a wind vane and a bottle barometer, and using them in observing and recording the weather

Explaining the factors that affect evaporation

Arranging wet things to dry fastest and keeping moist things from drying out

Keeping accurate records of weather data and using them to make simple weather forecasts

Interpreting newspaper and TV weather maps and identifying major features such as fronts, air masses, and areas of precipitation

Using limited fluid pressure over an extensive area to apply great force to an object to be moved or supported

Using correctly terms such as *atmosphere, volume, pressure, expansion, barometric pressure, evaporation,* and *condensation*

Facts and Principles to Be Taught

Air is real stuff just as pebbles and water are.

Air can be squeezed into a small space, but the smaller the space, the more pressure it takes.

Air fills almost all common spaces that seem empty, and before another substance can fill these spaces, the air must come out.

Air tends to expand when it warms and to contract when it cools.

Like water or soil, air cannot be moved without effort; thus it slows things that move through it.

When air is compressed, its temperature rises; when it is allowed to expand, its temperature falls.

The atmosphere exerts a surprising amount of pressure on all the things it surrounds.

The pressure of a confined fluid is equal in all directions at any point within the fluid.

Because cold air cannot contain as much water vapor as warm air, condensation may occur when moist air cools.

A cloud is visible evidence of a process; it is continually changing and is not a permanent object in the same sense that a stone is.

The climate is most severe near the surface of the ground.

The effect of climate on humans is minimized by clothing and other protection that some plants and most other animals do not have.

Sets of weather conditions recur from time to time and can be recognized from past records.

The basis of most short-range weather forecasts is extrapolation from present patterns and conditions.

BAG OF WIND

It takes time and many experiences for young children to understand that air is real stuff like water, even though it cannot be seen. One interesting experience is to let each pupil catch and investigate some air in a plastic sandwich bag, as follows.

Hold open the mouth of the bag and swish it through the air to fill it. Quickly close the mouth and twist it tightly to trap the air inside. Squeeze the trapped air. *What are some words that tell what it feels like?* Hold the twisted part tightly and sit on the bag. *Can air hold you up?* If your bag breaks, fill another, but this time ask some classmates to put their bags, together with yours, under a large thin board or book. Sit on the board. *Can three or four bags of air hold you up better than one? How big a person do you think you could hold up if you had many strong bags of air? What if you had only one big bag?* Try it!

Look through your bag of air at classmates. *Can you see them? Can they see you?* Open your bag a little bit and smell the air. *Do you smell anything different from the bag itself?*

Open the bag wide, hold it by the corners of the closed end, and dump out the air. Twist the neck shut again, all the way to the bottom of the bag. *Is any air left in the bag? How can you tell?*

Again, fill the bag with air. Close the mouth of the bag and take it outdoors. Open the bag and empty out the air. Fill it with outdoor air and bring it inside. *Can you be sure that none of the air from the room is still in the bag?* Do all the things with the bag of outdoor air that you did with classroom air. *In what ways is outdoor air like classroom air? How is it different?*

Now refill the bag, and twist it tightly shut. Put a rubber band around the twisted part to hold in the air. Try to push the bag down into a pail half filled with water. *Does the air keep the bag puffed, even under water? Do you think you could push the bag down in a lake? What if it were a very big bag of air?*

EMPTY OR FULL?

To most young pupils, a tumbler or other container that holds nothing visible is empty. But some simple and entertaining activities will show that even though pupils cannot see anything in it, an "empty" container may be full.

Divide the class into a few groups, each of which has an aquarium or plastic pail two-thirds full of water, and a plastic tumbler. (Plastic will not scratch or crack the aquarium.) Give each pupil a drinking straw, and then let each group proceed as follows.

Ask one pupil to put the tumbler into the water so that it fills completely. While it is submerged, ask that it be turned bottom side up and leaned against the corner of the aquarium. It should be held so that the open end stays tilted. Another pupil should insert the end of a drinking straw under the tilted tumbler and blow a bubble. *Where does it go? Is there as much water in the tumbler as before? If not, what does the tumbler contain besides water?*

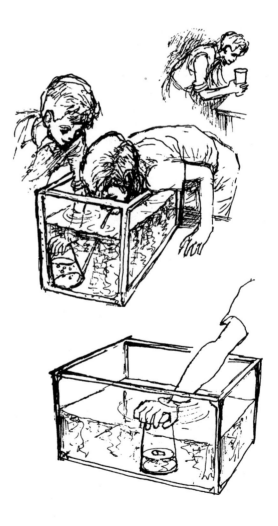

While one pupil continues to hold the tumbler in position, the others should take turns blowing bubbles until it is only half full of water. *Now what does the tumbler contain—half water and half what else?*

When all the pupils in a group have added enough breath to the tumbler so that all the water has left it, let the one holding it remove it from the water, still keeping it bottom side up, and set it on a table. *What is in the tumbler now? Is it empty or full?*

Ask that the tumbler be turned right side up, slowly. *Does it look the same as when it was empty? Even though it looks empty, what does it contain?* Let one pupil in each group put a small cork or a wood chip on the water. Then have that pupil invert the tumbler over the cork and push the tumbler down into the water so that the group can see how the tumbler's invisible contents keep the water from entering.

Ask the class to look for other full containers that seem empty. *What is in an "empty" wastebasket? Can anyone really empty a wastebasket? Can anyone find a really empty container?*

WEE BREEZES

Part of the experience of young pupils studying air should be to observe some ways in which air is made to move. They cannot see air, but they can see small, light objects move with it, and this helps them to visualize what the air is doing.

Give each small group of pupils a flat sheet of stiff cardboard about 11 by 17 inches (such as comes in a carton of duplicating paper) and a handful of puffed rice. (Puffed rice is easy to see, and easy to sweep up afterward.) Ask pupils to lay the cardboard on a desk or the floor and sprinkle a few grains of puffed rice around its edge. Then ask one of them to grip two edges of the cardboard and lift it quickly, keeping it level. *What happens to the puffed rice?* Let each pupil in the group try this several times. Then invite suggestions from all the pupils to explain why the puffed rice moved.

Ask the pupils to put the cardboard down again and sprinkle puffed rice around the edges as before. This time ask one pupil to raise, slowly, one end of the cardboard several inches and then let it go. *What happens to the puffed rice? Does the cardboard fall into* exactly *the same spot in which it lay before being raised?* Again, let each pupil have a try before inviting explanations.

Now sprinkle some puffed rice on the floor along the inside of a closed door. *What happens to the grains when the door is opened quickly?* Then, while it is open, replace the puffed rice along the threshold and quickly close the door. *Do the grains move the same way now as when the door was opened?*

In autumn, collect some fluff from seeds such as those of milkweed, cattail, or thistle. Put some on each desk along both sides of an aisle and ask a pupil to walk past them. *What happens to the fluff?* Put some around a sheet of paper, as puffed rice was put around the cardboard; then try lifting the paper. *Can this be done without disturbing the fluff? Can any movement be made in air without causing a breeze?*

This helps to show why air, even when it seems to be very still, is really in motion. Even a walking bug makes a wee breeze!

AIR PUSH

With a widemouth gallon jar, a strong plastic bag of similar size without leaks, and some string, a child can have truly fascinating experiences with the push of air. *Everyone* in the class should have these experiences; a demonstration would spoil the fun.

Gathering a complete set of materials for each individual requires some effort, but the items can also be used for other purposes. School cafeterias often can supply the jars, while children can bring the plastic bags and string from home.

To start with, ask the pupils to fill their bags with air. Usually they will do this by blowing, but show them that a bag can also be filled by pulling it through the air quickly, with its mouth held open. (See "Bag of Wind," page 30.)

Next, have them help each other tie an air-filled bag, upside down, to each jar—with its mouth over the mouth of the jar. They should wind a string *tightly* around the bag and jar several times, without crossing the ridges of glass, and tie it with a bow knot.

Then ask the children to press down on the bags, lean on them, and rest objects on top of them. *Why don't they go down? What other things act like this?* Pupils may mention inflated beach toys, air mattresses, and tires.

Now have the children untie the bags, put them down *inside* the jars—with the mouth of each bag folded over the mouth of the jar—and again tie them on *tightly*. When all are ready, ask them—*all at the same time*—to hold the jars and pull out the bags. Surprise!

The reason is that each bag acts somewhat like a hammock with a person lying in it. To pull up the hammock, one also has to lift the person resting on it. Likewise, to pull up the bag, one has to lift the air resting on it. This air extends as far up as air goes—hundreds of kilometers. No child can lift this much air; it weighs far too much!

STRAWS IN THE WIND

Unfortunately, some homemade wind vanes shown in science books would not work if tried. Because of poor design they would not point steadily into the wind, as they should. Such errors ought not occur. However, they may be turned to good advantage if used to make pupils aware that books are not infallible.

To this end, and also to help pupils develop creativity and the skill of analyzing how things work, let them design and make some simple wind vanes. These should be of their own invention, constructed from drinking straws, cardboard, heavy paper, pins, paper clips, pencils, rubber bands, and the like.

Let the pupils test their inventions outdoors, away from buildings, when there is a steady breeze blowing. *Which wind vanes turn into the wind, easily and steadily? Why do some turn more freely than others? What makes some stay level, while others keep one end low? Why do some point* with *the wind instead of* into *it? Why do some turn sideways to the wind?*

The pupils should change their inventions, if necessary, to make them work better. They should also make some of the wind vanes suggested in books, and test them. *Who can tell, even before trying these, what must be done to improve them?*

A wind vane must point into the wind and readily show changes in wind direction. It will—if:

1 It can turn easily on a pivot, and remains level regardless of which way it points.

2 Its tail extends farther from the pivot than the head does, to enable a force to have a greater turning effect on the tail of the wind vane than on the head. (See ''Paper-Clip Cranks,'' page 121.)

3 Its tail has a larger area than the head, so that the wind exerts more force on the tail than on the head.

4 Its head is weighted in order to balance the larger tail.

5 Its tail is in two parts, spread slightly, to steady the wind vane.

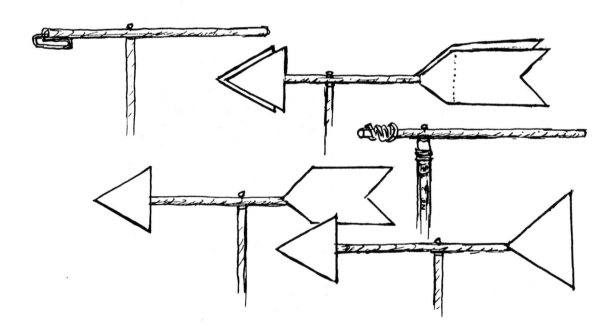

BLOWING UP THE TEACHER

Air under pressure is used to operate many devices that require great force. Car lifts in service stations use compressed air. Air brakes on a truck apply far more force than drivers can with just muscles. And air-filled tires support cars and trucks.

Because air is a fluid, it can enter small openings and fill spaces of all shapes easily. Also, the pressure at any point in confined air is the same. Thus, air can be squeezed into a tire through the tiny opening in the valve stem. Once in the tire, air presses on all parts of the tire equally. The larger the area on which the confined air presses, the greater is its total force.

It is fun and exciting to combine air under pressure and a large area as follows. Get two large tables that are identical in surface area. (Folding tables about 1 by 2 meters in size work well.) Or take the class to a room where there are two large identical tables. Have as many children as can fit around one table kneel at its edge. Give each child a gallon-size plastic bag. Have each one pucker the mouth of the bag as if to blow into it, but lay the bag flat on the table with just the mouth projecting.

Then, while the children move away for a moment, let others help in turning the second table upside down over the first. Only the puckered mouths of the plastic bags should protrude from between the tables. Now have the children return to their places at the tables' edge.

Now ask a volunteer to get up on the overturned table. (The pupils will be even more impressed if *you* volunteer.) Ask the children not to blow into the bags until you give the signal. Then say, "One . . . two . . . three . . . Blow!" When they all blow simultaneously, the table with you on it will rise a short distance. The children will have blown up the teacher!

This dramatic display of great force resulting from air pressure applied over a large area can be put to use in a number of ways. Let the class try to move some heavy objects by means of air pressure. A filing cabinet held away from the wall by a baseboard may provide enough space for inserting a number of plastic bags. *Can the class move one by applying "lung power" to the back of the cabinet?* Perhaps someone can bring in an air mattress. Then, while one person lies on the deflated mattress, another can blow into it. *Can one person lift another with lung power?*

The pressure of the wind on a windy day is much less than lung power. *Why, then, must a builder brace a long concrete block wall when a building is under construction?*

BOTTLE BAROMETER

Because of its weight, the air around us presses against everything it touches. Its pressure often changes. This can easily be shown by a simple air-pressure indicator.

To make such an indicator, find a bottle with a long slender neck. Invert it in a jar of water. Then warm it with both hands until several large bubbles of air escape. *What causes this to happen? When the bottle is allowed to cool again, why does water rise in its neck?* (See "Shrinking and Swelling Air," page 178.)

Stand the jar and bottle in a place where the temperature is steady—perhaps inside a closed cabinet. Place a thermometer near them. Then let pupils check from time to time. They will see that the water level in the neck of the bottle changes—even if the temperature does not.

This is because of changes in air pressure. There is a sort of contest—a "push-of-war"—between the air inside the bottle and the air outside it. The air inside pushes against the water in the neck of the bottle, and tends to make it go down. The air outside the bottle pushes against the water in the jar, and tends to make it go up into the neck of the bottle. (The bottle does not fit tightly on the jar.)

If the pressure of the air outside the bottle becomes greater, what will happen to the water level in the neck of the bottle? What if the pressure of the air outside the bottle becomes less?

For keeping a record of the changes, a scale is needed. This may be a large index card with its lines numbered, taped to the jar so that the lines face inward. With this scale, the indicator is an air-pressure gauge or meter—a *barometer*.

Pupils can then record, from time to time, the position of the water level in the neck of the bottle. *Why should this be done only when the thermometer shows a certain temperature—the same each time—such as 20°C? What might happen to the water level if the air inside the bottle became warmer? Cooler?*

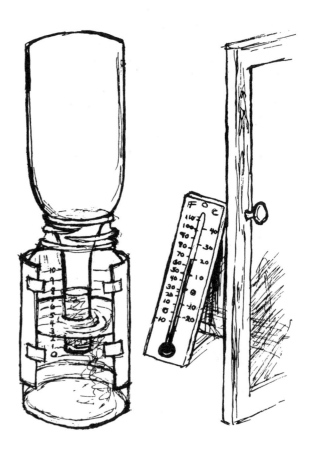

ONE-GALLON CLOUD

As a sequel to "Air Push" (page 33), pupils can easily make a miniature cloud. They need only set up conditions similar to those which often cause clouds to form in the sky.

It is best to try this yourself, first. Use the same widemouth gallon jar, plastic bag, and string that were used in "Air Push." However, a sheet of rubber obtained from a large round balloon by cutting off its neck, and several stout rubber bands, may work even better than the plastic bag and string.

Start by pouring a little water into the jar. Then light a match, blow it out, and hold it inside the jar to introduce a little smoke. Next, put the plastic bag down inside the jar and tie its mouth *tightly* over the mouth of the jar. Or stretch the sheet of rubber from the balloon over the mouth of the jar and fasten it tightly with the rubber bands.

Then, while someone holds the jar firmly, pull up on the plastic or rubber—*hard*! This will cause a cloud to form inside the jar, best seen if bright light shines through the jar. The cloud will disappear when you stop pulling.

Now let pupils try this, and also carry on some investigations. *Can a cloud be made without water in the jar? With water, but without smoke? Do other impurities, such as automobile exhaust or frying-pan smoke, work in place of match smoke?*

A cloud forms in the jar because of a chain of events. The air inside the jar contains water vapor, which is invisible. This air is pressed upon by the air outside, even though the plastic or rubber separates them. However, the pressure on the air inside the jar becomes less when one pulls up on the plastic or rubber. As a result, the inside air expands, and in doing so, becomes cooler. And, as it cools, some of the water vapor in it changes to liquid water around the particles of smoke, forming tiny droplets. Thousands of these droplets are what make up the cloud.

Outdoors, most kinds of clouds, especially *cumulus* clouds, form in much the same way. When air moves to where there is less pressure on it—as it does when it rises—it expands and cools. Then some of the water vapor in it may condense on minute particles of impurities, forming a cloud of tiny droplets. Not all clouds, however, are formed like this, and not all consist of water droplets. Some, called *cirrus* clouds, are made of ice crystals.

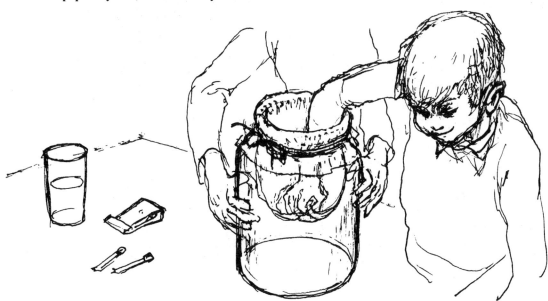

THUNDERSTORM PATHS

To many persons thunderstorms are noisy, awesome, and often frightening things that appear and disappear rather unpredictably. But these storms often move along in paths that your pupils can plot at school or at home. Then the confidence that comes with measurement may help to dispel the fear that is fostered by ignorance.

Have the class prepare plotting maps in advance, using large sheets of paper and drawing compasses. With the compass point at the center of each sheet, let them draw a series of concentric circles. Starting with a circle of 1 centimeter radius, each successive circle should be 1 centimeter larger in radius, to 15 centimeters or so. The circles should be labeled "1 kilometer," "2 kilometers," "3 kilometers," and so on, to represent the distance from each pupil's position at the center of his or her sheet.

Next, have the pupils sketch the position of major land features on the map, including hills, large buildings, and waterways. As on most maps, north should be at the top of the sheet.

When thunder is heard, have the class move to a place where they can watch for lightning. Each map should be oriented so the north edge is toward the north. Each time they see a lightning flash, the pupils should begin to count seconds (having practiced using the "Swinging Second-Timer" described on page 19) until they hear the thunder. Since sound travels about 370 meters per second in warm air, each three seconds that elapse between lightning and thunder means about a kilometer of distance. From the flash, the pupils know the direction to the storm. From the delay in the arrival of the thunder, they can tell how far away it is.

For each lightning flash, and subsequent clap of thunder, each map marker should make a dot on his or her map, labeling it "1" if it is the first observation, "2" if it is the second, and so on. Cloud-to-ground strikes will be easier to plot than cloud-to-cloud flashes.

Do the plotted points lie in a line, or do they skip around? How fast does the storm move, and in what direction, according to the scale of the map? How wide is the path of the thunderstorm? Occasionally, the plotted path may show its direction to be directly toward the observers. Then it is interesting to try to pinpoint the arrival time of the storm, taking cover before it happens. *How close is the forecast made from such a map?*

Do not be surprised if curiosity and questions replace awe and fear on the part of pupils who try this activity!

RAINDROP RECORDS

How big do raindrops get? From what kinds of clouds do the largest ones come? These are questions that might be asked by children. And answers may seem difficult to find, since raindrops splash when they hit. But there is a way to record their size that is both fun and instructive, as follows.

Get some old pantyhose. Also, get an embroidery hoop, some powdered sugar, and a pie tin at least as large as the hoop. Sprinkle some powdered sugar in the pie tin. Then cut a piece from the pantyhose large enough to be stretched across the opening of the embroidery hoop. Dip the stretched fabric in the powdered sugar to coat the mesh. Tap the hoop to knock off any excess sugar. (The sugar that sticks does so, in part, because of an invisible coating of skin oil that is left on the mesh. Washing removes the coating and makes it harder for the sugar to stick.) Set the hoop with its coated mesh aside until a rainy day.

During a rainstorm, take the hoop and mesh outdoors, holding it under an umbrella until you are ready to make some raindrop records. Then, while you count off seconds, hold the hoop horizontally in the rain for a few seconds, letting the raindrops strike the sugar-coated mesh. Return it to the protection of the umbrella, record the exposure time, and return to the classroom.

Indoors, examine the fabric for spots left where the raindrops passed through the mesh, removing the sugar as they went. These are records that you can measure, and even preserve if the remaining sugar is not knocked off the fabric.

Some things to measure, calculate, and think about, using the raindrop records, are:

1 *What is the diameter of the largest raindrop that you recorded?*

2 *What is the* average *size of the raindrops that fell on the mesh?*

3 *From the number and diameter of the raindrops that left a record, what* volume *of water fell on the mesh during the time it was exposed to the rain?* (Hint: $V = \frac{1}{3} \pi r^3$)

4 *From the area of the hoop, the time of exposure to the rain, and the mass of the water you calculated, how much water would have fallen on the area of the hoop in an hour? At that rate, how much water would have fallen on the school lawn in an hour?*

RAINFALL MAGNIFIER

Pupils are generally aware of rain, because it affects their activities. However, they probably do not appreciate how varied the amount of rainfall can be. Some simple investigations of the depth of rainfall in containers will help to make them aware of these variations, and of the wind direction, cloud forms, and pressure changes associated with rainy weather.

At the start, ask each pupil to bring to school a straight-sided, topless container. See that the assortment includes metal cans, jars, and milk cartons, as well as a sand pail or plastic bucket with sloping sides and possibly a cone-shaped beverage container. Let several of the class cut "empty" (see "Empty or Full," page 31) cartons to half their normal height, each using one of these for his or her container. The rest should use containers from the assortment brought to school. Ask the pupils which of all the containers they think will collect the deepest water, and which will collect the greatest volume during a rain. Let them try to rank the containers in this way, giving reasons.

On a rainy day, let each pupil carry his or her container, inverted, outdoors to an unprotected place. At a signal, the containers should be set right side up on the ground to catch the rain. When the rain has stopped, or at the end of a particular time, such as an hour, the class should collect its containers for examination. As a signal was given for the rain-catch to start, a signal can be given for it to stop and the containers covered so that the same time elapses for rain to fall into each.

In the classroom, each pupil should use a ruler to measure the depth of water in his or her container. *Which shape has the deepest water? Which has the shallowest? Do all similar containers (milk cartons, for example) have the same depth of water? What about other straight-sided containers? Does the container with the deepest water also contain the greatest volume of water?*

To answer this last question, let each pupil, in turn, pour the rain water from his or her container into a straight-sided olive bottle, marking the water level on a strip of adhesive tape fastened to the side. *Which container did collect the greatest volume of water? Which collected the least? Why is the rain water deeper in the olive bottle than in the container that collected it?*

Since most rainstorms drop only a centimeter or so of water, it is easy to make large errors in its measurement. An observer who makes an error of only 1 millimeter in measuring 5 millimeters of rain water would make a 20 percent error (1 millimeter in 5). If there were 50 millimeters of rain water, however, and the observer made the same error of 1 millimeter in measuring its depth, the error would be much smaller (1 millimeter in 50, or only 2 percent). For this reason, a commercial rain gauge usually collects rain from a large-diameter circle and directs it into a smaller-diameter container for measuring.

Now show each pupil how to make a simple rain gauge as follows. Use an olive bottle or small

juice can, and a large juice can whose top has been removed. First, tape two straight pins across the flat side of a ruler so that the point of one projects exactly 1 centimeter farther from the edge than the other. Fill the large can with water until the level comes just to the upper pinpoint, with the edge of the ruler resting across the mouth of the can. Then transfer water from the can to the bottle with a spoon or medicine dropper until the level drops to the lower pinpoint. On a strip of tape fastened to the outside of the bottle, mark ''1 centimeter'' where the water level comes. Repeat the process, adding another centimeter of water and marking ''2 centimeters'' on the tape. *Where would ''½ centimeter'' and ''1½ centimeters'' be marked on the tape?*

To make a rain gauge that does not depend upon pouring water from a can into a bottle, set a funnel into the top of a large can to direct the rainfall into a smaller one under the funnel's spout. The magnifying effect of such a gauge is the *ratio* of the area at the funnel's rim compared with that of the smaller can. (It is also the ratio of the *squares* of the funnel rim's diameter and the small can's diameter.)

When the rain gauges have been completed, let the pupils set them outdoors in unprotected places to measure the rainfall in several storms. Some questions to investigate with their rain gauges are:

1 *How much rainfall does a thunderstorm deliver?*

2 *Which gives more rainfall—a thunderstorm or an all-day drizzle?*

3 *What combination of wind direction, air pressure, and cloud forms produces the most rainfall?*

4 *How does the rainfall vary with distance from the trunk of* (a) *a deciduous tree?* (b) *a coniferous tree?*

Plants and Animals

The interest of children in plants and animals is obvious. Their passion for pets, their frequent imitation of animal sounds, and their irresistible urge to pick a pretty, fragrant flower—all attest to their fascination with living things. Properly nurtured, this interest should continue for life, maturing to include an appreciation for once-thought "ugly" animals, plants that have no flowers at all, and the interdependence of organisms.

Plants and animals can be studied at any season, indoors or out. Even in the "dead" of winter, trees are very much alive. So are seeds on weed stems and in packets, creatures in ponds, and spiders in cellars. The pupils themselves—the most important organisms of all—are available for study.

The study of plants and animals requires little more than keen eyes and an inquisitive nature. Magnifiers and microscopes are helpful but not necessary. Materials are free or inexpensive, and there are plants and animals near every school.

Pupils can experiment freely with plants and not feel that they are hurting them. However, observations or experiments that may cause injury or distress to animals should be discouraged. There is plenty to do and to observe without risking an animal's well-being. Also, native animals are preferable to exotic species for classroom studies, for they can be released when interest in them wanes. Exotic animals cannot be released without possible harm to them or to our native species.

The lessons learned from studies of plants and animals are basic to desirable environmental ethics and practices. As people make increasing demands on living things for food, fiber, and recreation, they must make wise decisions to stem the rates at which living things are used and their ranges restricted. Also, if people are to live in harmony with each other, they must learn to respect all forms of life. Early experiences with plants and animals can help to develop such a respect.

SOME IMPORTANT OBJECTIVES

Attitudes and Appreciations to Be Encouraged

Basic to a person's respect for fellow human beings is a respect for *all* living things, no matter how lowly or insignificant they seem.

No creature is, in itself, harmful or beneficial, but only in terms of how it affects human beings.

Many creatures that at first seem uninteresting or even repulsive become fascinating and even appealing as they become familiar.

We are organisms whose bodies are marvelous machines that can be studied without special equipment and without physical harm.

Since nearly all plants and animals are interdependent with one another and with their environment, the consequences of a major action on any one of them must be thoughtfully considered.

Decay after death is not unfortunate and offensive; it is essential in order to recycle the substances needed for the living.

There are so many kinds of living things and they live in so many kinds of places that anyone who is interested can find out something new and exciting about them.

Skills and Habits to Be Developed

Noticing plants and animals, and animal traces, that might escape the eye of a less informed observer

Relating the age and growth of a felled tree to the appearance of its stump, and inferring the position of pieces of wood in the logs from which they were cut

Setting up and maintaining an aquarium containing a few organisms borrowed from a pond, and returning the borrowed organisms in time

Holding a hand magnifier and a small object to optimize both focus and lighting

Recognizing the factors that affect decay, and modifying these to speed or slow decay, depending upon circumstances

Setting up a controlled experiment with living things, and using a sufficient number of individuals to draw valid conclusions

Taking a random sample of a population and using it to count or to describe the whole

Working with others in carrying out an investigation and in gathering and analyzing data cooperatively

Holding an animal such as a harmless snake or a young bird so as not to cause undue discomfort or injury

Using correctly such terms as *insect* (versus *bug*), *fruit* (versus *seed*), *larva, sample, community, population, overproduction, decay, scavenger, environment,* and *spore*

Facts and Principles to Be Taught

Human beings are organisms whose skeletons, muscles, eyes, and other body parts are in many ways similar to those of dogs or cats.

Most muscles of backboned animals work in opposing pairs, one muscle moving a body part in one direction, and the other returning it.

All aquatic animals need oxygen. Some get it by coming to the surface; others have gills and get it directly from the water.

For every organism that is plainly visible, there are many others that go unnoticed because they live out of sight, are very small, or are in a stage that is hidden, protected, or motionless.

Green plants are the basic food makers for all other organisms. They are the first link in any food chain.

All organisms need oxygen to live, but green plants are the only ones that produce more than they use.

By producing far more offspring than can mature, most living things perpetuate themselves in spite of unfavorable conditions or calamities.

All plants and animals produce young, and they do so in a variety of ways—mostly by seeds, spores, or eggs, but also by new growth directly from the body of an adult.

Woody plants grow in length only near their tips, but they grow in diameter throughout their stems and roots.

Besides needing moisture and warmth, molds need food on which to grow, because they cannot make their own food as green plants do.

Decay is hastened not only by bacteria and molds but also by many small animals whose activity often goes unnoticed.

Spiders are one of our most numerous animals, have predictable habits, and indirectly may be of great help to people.

THE BETTER TO SEE WITH

The eyes of human beings face forward. Animals such as rabbits, robins, and goldfish have their eyes at the side. *Which are better to see with?*

Let the pupils choose partners and play catch with Ping-Pong, Styrofoam, or other light balls. The teams should record each time they catch the ball cleanly and each time they fumble or drop it. At a signal, all the pupils should close, or blindfold, one eye. Then they should continue to play catch, still keeping a record of catches and of fumbles or drops. *How does the record of catches using both eyes compare with the record of catches using only one eye? Why are there almost no one-eyed players in major league baseball?*

Have each pupil hold a hand over one eye and look at some object in the room. Then have each one open that eye and hold a hand over the other, alternating in rapid succession. *Is the scene exactly the same when viewed with either eye? Which objects seem to shift the most—nearby ones or distant ones?*

Have each pupil keep one eye closed and look at the same scene as before, while blinking the other eye. *Does the scene shift as before?* A person sees a slightly different view with each eye. This difference helps us tell how far away various objects are.

While the pupils keep looking straight ahead, have them stretch their arms out to the side and wiggle their fingers. *Without turning eyes or head, how far back can the pupils move their fingers before they disappear from sight?*

Now have pupils look at rabbits, robins, chickens, and goldfish, or at pictures of these animals. *Where are their eyes located in comparison with ours? Would a rabbit see most things with both eyes, or with only one? Does this help it tell distance? How far back do you think a rabbit could see without turning its head? Would it be easier to sneak up on a rabbit or a person? If you had a choice, would you rather see like a rabbit, or like a person? Why?*

Insects have an entirely different kind of eye. The eyes of the completely harmless dragonfly, for example, are composed of hundreds, or even thousands, of simple eyes, each "staring" in a different direction. Perhaps someone in your class can bring in a dragonfly for study with a magnifier. Multidirectional eyes help a dragonfly follow and catch small flies in midair!

There is much to see with eyes. But there is much to see *in* eyes, too!

MUSCLE TEAMS

Nearly all muscles work in pairs. One of the pair pulls a part of the body in one direction; the other pulls it in the opposite direction. Muscles act by becoming shorter, but they cannot expand by themselves. When one contracts and then relaxes, it needs a partner to pull it back. Pupils can learn about the teamwork of muscles by examining some muscles in their own bodies, as follows.

While seated at your desk, place your right hand under the desk and lift up. At the same time, use your left hand to feel the muscles of your right arm, above your elbow. Where the muscles are hard, they have contracted. Where they are soft, they are relaxed and long. *Which muscles are trying to lift your forearm?*

Now move your right hand, palm up, to the top of the desk and press down. Feel the muscles of your upper arm as before. *Which ones are contracted? Which are relaxed? Do they feel the same as they did when you were lifting?*

Next, rest your foot flat on the floor in front of you, pressing it down hard enough so it will not slide easily. With one hand, feel the muscles at the front (top) of your thigh and with the other the muscles at the back (bottom) of your thigh. Try to slide your foot away from you. *Which muscles contract? Which are relaxed?*

Try pulling your foot toward you. *What change can you feel in the muscles?* Push and pull your foot back and forth several times and note how the muscles alternately contract and relax.

Try other movements of your body and see if you can find opposing sets of muscles. *Can you feel the muscles that close your jaws? What about the ones that open them?*

Some muscle teams are not so easy to find. *If muscles can only contract, where are the ones that enable you to stick out your tongue? How can muscles make a worm long and slender?* (Hint: Try rolling a ball of clay into a long, thin "worm.") *Where do you think the muscles are that cause it to get this shape?*

THE LAWS OF JAWS

When pupils use the thumb and fingers of their hands to imitate the action of their jaws or those of a dog, they are apt to make both parts move up and down. If they investigate how their own jaws and those of other backboned animals really work, they may be surprised to learn that their hands did not show how jaws work at all.

Pass a cracker to each pupil. Half of it should be put in the mouth to chew while the pupil rests his or her *nose* on the edge of the desk. *Is it easy or difficult to chew like this? Is it natural for the jaws to move this way?*

After the pupils swallow the first half of the cracker, ask them to chew the second half. But this time each pupil should rest his or her *chin* on the desk while chewing. *Is it easy or difficult to chew now? When your lower jaw cannot move, what must your head do to chew?*

Ask the pupils to observe other vertebrate animals (ones with backbones), such as dogs, cats, horses, cows, fish, turtles, and birds. *In all these animals, which jaw moves in chewing—the upper, the lower, or both? Do the jaws of an alligator, when feeding, work the same way as our own?*

A snake's jaws are different from ours, making it possible for the snake to swallow prey larger than its own head. There are elastic connections between its upper and lower jaws that let the jaws separate. Also, the right and left halves of the lower jaw can be spread apart to help in swallowing. Try to have your class watch a snake as it swallows something big. *In what ways do its jaws operate differently from ours? Similarly to ours?*

Now ask each pupil to bring in a grasshopper or a beetle, both of which are chewing insects, and look carefully through a magnifier at the insect's jaws. *In what ways are they similar to our own? In what ways are they different?*

LAYERS IN LOGS

Good teaching is not simply giving information. It is also having pupils learn by observing things for themselves—tree stumps and logs, for example. For just this purpose it is highly desirable that some tree stumps be left on the school grounds!

Let pupils examine a stump and count the rings. *If one ring was formed each year, how old was the tree?* To help in counting, a pin bearing a small paper "flag" may be stuck into every tenth ring. A magnifying glass will also help.

How big around was the tree when you were born? How big was it when it was as old as you are now? Let pupils outline these sizes with crayons. *Was the tree living when Lincoln was President, or when Victoria was Queen? If not, can you find the stump of a tree that was living then?*

The rings seen in tree stumps and logs are really the cut edges of *layers* of wood, a new one formed each year, surrounding all the rest. These layers nest inside each other—a little like upside-down ice-cream cones in a stack. However, they are much longer than ice-cream cones, and taper much more gradually. They are also far more irregular, with many extensions into branches and twigs.

When possible, have pupils count the layers in several sections of a cut-up tree trunk and branches. *Why is the number of layers so different from place to place? What may have caused the differences in thickness of the layers?*

Also let pupils examine scrap pieces of wood for layers, as well as solid wood furniture, baseball bats, and even pencils. *How long did each piece of wood take to grow? In what position must it have been in the log, before sawing?*

Can you find log layers in plywood? Pupils may read about how veneer and plies for plywood are made—by "peeling" continuous, thin sheets of wood from logs. Cut in this way, the layers in the logs cause the patterns often seen in plywood.

How could you prove *that each layer in a tree represents one year's growth? How might you find out why the layers are not of uniform thickness?* Reading about a tool called an *increment borer* may help!

LIMB LIFT

When a pupil grows, he or she grows all over. *Do woody plants grow in the same manner?* Some interesting observations and activities will help to provide the answer, and will change opinions, too.

First, ask the pupils if tree limbs, like their own arms, get lifted higher as a tree grows. Record their opinions. Then set up the following committees to study the question:

Swing   Seedling
Trunk   Fence
Vine

Ask the Swing Committee to investigate, both by interview and direct observation, whether swings hanging from limbs must be lengthened from time to time as the trees grow.

Ask the Fence Committee to look for fence wires that have been fastened to tree trunks. *Are old fences higher than new ones? Why do some fence wires or chains seem to go through a tree instead of being attached to the side? Do farmers have to lower fences from time to time as trees grow?*

The Vine Committee should examine how vines fasten themselves on buildings and other structures. *Do vines move their stems as they grow, refastening them to higher places?* Mark a vine and the wall behind it at the same level. *Do the marks move apart in the course of a week, or a month, during the growing season?*

The Trunk Committee might try sticking two pins, one a meter above the other, in each of several tree trunks. Each day, for several weeks during the growing season, this committee can measure the distance between the pins. *How much does the distance change during this time?*

The Seedling Committee should make ink marks 1 centimeter apart on the stem of a seedling such as a lima bean plant 10 to 20 centimeters tall. Then the committee can keep a record of the spacing of the marks as the plant grows. *Do the marks all move apart at the same rate? Where must new marks be made?*

When sufficient evidence seems to have been collected, let the committees report their findings in a plenary session. *What parts of woody plants grow fastest, according to the evidence? Do woody plants really grow as people do?*

SEED SURPLUS

An important biological principle is that most plants and animals produce far more offspring than are able to mature. To make this idea more meaningful to pupils, let them discover how many seeds a single plant can produce. Many common weeds work well for this, before their losing seeds.

First bring in a large weed with seeds, and ask the pupils to write down their estimates of the number. Then have them collect all the seeds and determine the number more accurately. They may count out groups of 10 or 100. Or they may divide the seeds into piles of equal size, perhaps thimblefuls, count those in a few piles to get an average, and multiply this by the number of piles. Otherwise, with a sensitive balance they can find out how many seeds it takes to weigh as much as a paper staple or paper clip and then how many staple weights or paper-clip weights in the entire lot. (See ''Three-Corner Balance,'' page 27.)

It is not good science to generalize on the basis of *one* sample. Therefore, let each pupil find the number of seeds on a separate plant of the same kind, chosen at *random*—not just the biggest one that can be found. A class can get a random sampling by spreading out across a weedy area and then picking the nearest specimen of a particular kind of weed when you call, ''Stop!'' All the individual estimates can then be used to find the *average* number of seeds produced by plants of that species in that particular place and year. A class can do this for several species, or committees may be responsible for different kinds of weeds.

What would happen if all these seeds grew into mature plants and the mature plants all produced seeds, all of which grew, and so on? Why would all the seeds not *be likely to grow into mature plants? Would some have a better chance than others? If so, which ones? Why? How many seeds actually do grow when planted?* Pupils can find out by planting 100 in a pan of soil.

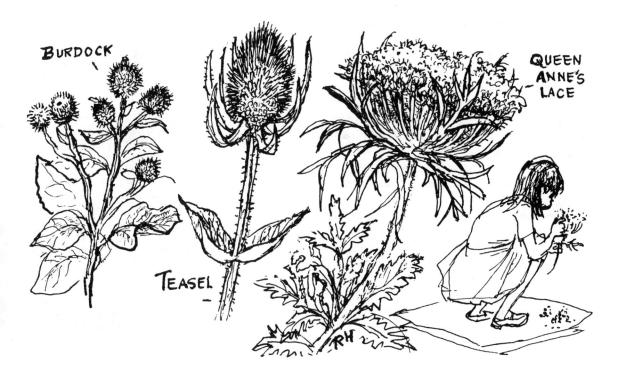

BURDOCK

TEASEL

QUEEN ANNE'S LACE

FISH-LESS AQUARIUMS

Pint jars make excellent aquariums for pond life— if fish and other large animals are left out. *Every pupil should have one of his or her own to watch and enjoy, even if the windowsills and book-shelves are filled for a time!*

Such aquariums encourage pupils to make observations on their own and to share their discoveries with classmates. This is so much better than merely feeding goldfish in a classroom aquarium that has become just part of the furniture!.

To make a fish-less aquarium, a pupil should place a centimeter or two of pond-bottom sediment in a jar, stick in a small water plant or two, and add pond water. Then in may go a snail or two and a *few* water insects. The water will become clear after a while, allowing many tiny animals to be seen, perhaps for the first time. A magnifying glass will help.

Ideally, a class should set up jar aquariums at a nearby pond—borrowing, in a sense, small parts of the pond for a little while. Early fall is a good time to do this. Then the experience may be repeated in the spring, to see what seasonal changes have occurred.

Well in advance of the trip to a pond, notes should be sent home to advise about clothing, especially rubbers or boots. Also, one or two parents may be invited along to participate, as well as to provide additional supervision. Possibly a primary-grade teacher and an upper-grade teacher can take their classes at the same time, having each older pupil be responsible for a younger one. It is surprising how well they shoulder this responsibility!

If the whole class cannot be taken to a pond, perhaps you and a few pupils can bring the materials to school. For a class of 30 pupils, you should fill four gallon jugs with pond water and get a small pail of sand or silt from the pond bottom. Also borrow some plants and animals from the pond—*not too many!* Gather a few handfuls of small plants, and use a large kitchen strainer to collect some water insects, snails, and other small animals. Put them in a gallon jar of water.

If each sample is first dumped into water in a white enameled pan, the creatures will be more easily seen. Then they can be spooned out. At school, they may be transferred to shallow enameled pans, for pupils to select the ones they want.

To make the experience a really pleasant one, be sure that pupils:

Put only a *few* animals and plants in a jar—too few rather than too many. Crowding may cause them to die.

Keep the jars in a cool place, away from radiators. As a rule, they should not be left in direct sunlight.

Move or shake the jars as little as possible. The creatures that live in them should be observed, not disturbed.

Add *no* food. The green plants make their own food, and the animals eat plants, or animals whose food came from plants.

Use pond water to replace any water that evaporates. Tap water containing chlorine is harmful to some creatures.

Over several days, pupils may investigate questions such as these:

1 *How do the animals get around—by legs, or by other means? Do any stay in one place, attached to objects?*

2 *What do the animals eat, and how do they eat? How do they keep from being eaten?*

3 *Which animals carry bubbles of air with them, or get air through little tubes? Do any get air by moving gills through the water?*

4 *Do some animals avoid light? If you leave a paper bag over a jar, and later remove it, can you see things you did not see before?*

5 *Where do the plants show signs of most growth—in leaves, stems, or roots? Do they all have these parts?*

Among the many other things that pupils may discover are these:

Hydra, named after a monster of mythology, snaring other small animals with their "arms"

Tiny water "fleas," such as *Daphnia* and *Cyclops,* darting through the water

Small shrimp-like animals with curved backs and many legs, called *scuds,* swimming along

Diving beetles carrying their own air supply with them

Caddis larvae pulling their cases along

Mosquito larvae or "wigglers" hanging from the water surface, later changing to mosquitoes

Embryo snails developing inside jellylike snail eggs, and later hatching

Anacharis (once called *Elodea*) plants giving off tiny streams of bubbles in sunlight

After a week or so, before interest wanes, have pupils return the plants and animals to the community from which they were borrowed. Take another trip to the pond—make it *just as important* as the first one—for the express purpose of putting the creatures back. Doing this, instead of flushing them down the drain or throwing them out, helps to instill a respect for living things—a basic environmental ethic!

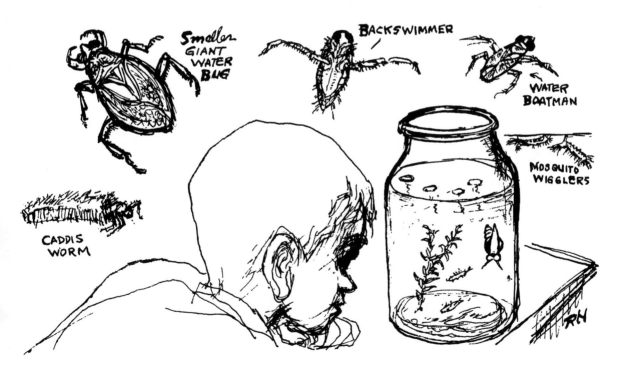

Smaller GIANT WATER BUG
BACKSWIMMER
WATER BOATMAN
MOSQUITO WIGGLERS
CADDIS WORM

GREEN FACTORY

All living things use oxygen. By the process of photosynthesis, however, green plants produce more than they use. They constantly put back into the air oxygen that is removed by animals and by the burning of fuels, rusting, and other sorts of oxidation. When children look at trees and lawns, they may be aware that photosynthesis is going on. *But do they—or you—have any idea of how vast an area of leaf surface is involved?* To find out, let the class take some "green factory" measurements as follows.

Find the dimensions of the school lawn and outline it to scale on construction paper, oak tag, or cardboard. Cut out this outline and find its mass. (See "Calibration and Confidence," page 24.) Next, from the same material cut out a scale *unit of area* (such as a square representing 100 square meters of lawn) and find its mass. From the *ratio* of the masses, you can determine the area of the school lawn, no matter what its shape.

Now select a small sample of grass, perhaps a square 5 centimeters on a side. (*What fraction of a square meter would this be? What fraction of the school lawn?*) Cut the grass and the leaves of other plants in this sample area and bring them into the classroom. On a sheet of construction paper, trace the outline of all the leaves. Cut out the outlines and find their total mass. Then find the mass of a unit area of the same material, such as a square 10 centimeters on a side. *From the ratio of the masses, what is the area of the leaves that you cut? If that is the area of leaves for 25 square centimeters (5 centimeters on a side), what would be the leaf area of the whole lawn? About how much green factory area is removed when the lawn is mowed?* Fortunately, it grows back!

The same method of measurement may be applied to a tree. First, without breaking off branches, try to estimate how many leaves there are on a mature tree. Do this by counting the leaves on a few *representative* branches. Then multiply this figure by what seems to be the number of such branches on the tree. Next, get a representative leaf from the tree and trace its outline on cardboard. Cut out the outline and find its mass as before. Compare this with the mass of a unit area of the same material. *How many leaves would it take to total 1 square meter? About how many leaves on the entire tree? Therefore, how many square meters of leaf area on the whole tree?*

It is interesting to compare the leaf area of plants with the area of soil directly beneath them. *How does the leaf area of a tree compare with the area of soil directly beneath the tree? What do you think would be the area of leaves in 1 acre of corn? How could you find out?*

STICKY TRAPS

Many living things can be studied in the class-room, but some can be studied better where they normally live. Spiders and their webs, for exam-ple, are observed best in gardens, in shrubbery, or in dark corners of buildings where things are un-disturbed. To help your pupils appreciate what an amazing structure a spider's web is, suggest that they investigate some webs as follows.

Look for an orb (round) web that is nearly per-fect, indicating that it was recently made. *Is the web more nearly horizontal or vertical? Which would be better for catching flying insects? On what part of the web does the spider rest?* If it is not on the web, see if you can find its hiding place. *Can you find how it can tell when something is caught in the web?*

With a toothpick, lightly touch each kind of silk making up the web—the main supporting lines, the radial (spokelike) lines, and the apparently cir-cular ones. *Which are sticky enough to hold a fly?* Follow one of the apparently circular lines. *Does*

it join itself to make a circle, or does it form part of a spiral? If you can catch a fly, toss it into the sticky threads; then watch to see how the spider approaches it and wraps it in silk.

Now look for a funnel web. (*Note:* The funnel-web spiders of North America are harmless. Only tropical ones, such as the funnel-web spider of Australia, should be avoided.) Often funnel webs are hidden in bushes, tall grass, or in dark corners of buildings, so you may need a flashlight to ex-amine one. *Is a funnel web sticky like an orb web?*

The funnel-web spider usually waits in the throat of its funnel and then darts out after prey that alights on the web. Drop a fly or a grasshopper on the web and see. Probe downward into the rear of the funnel and see if you can make its occupant come out. *Does the occupant walk or run out? Does it have the same number of legs as other kinds of spiders you have observed?*

Finally, look for a spider lowering itself on a single thread. *Is the thread sticky? If you can make the spider scramble back up, what does it do with the thread?*

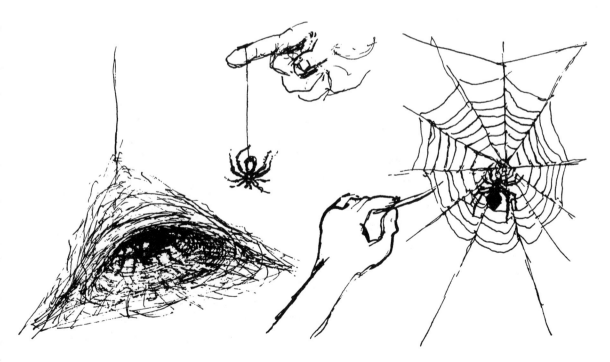

LIVING FUZZ

To most pupils the word *plant* may suggest "green," "seeds," or "flowers"; however, there are many plants that are not green, do not come from seeds, and do not have flowers. These are interesting to study and some are easy to grow. Molds are some such plants.

Ask the pupils to bring in some fruits such as berries, grapes, or plums, and some samples of home-baked goods such as bread, cake, or rolls. (Store bread usually has a chemical to inhibit the growth of molds.) In class, give each pupil two plastic sandwich bags. Have each write his or her name on two slips of paper and put one in each bag.

Next, have each pupil moisten *slightly* a piece of bread or cake and put it in one bag. In the other bag have the pupil put a sample of fruit. Then have all twist their bags shut and fasten them with rubber bands. Let each row or group of pupils put their bags in a carton. Then set the cartons in a warm, dark place.

In three days, or after a weekend, let the pupils remove the bags and, without opening them, examine the contents. *Has any fuzz appeared? Does it appear on just the food, or on the plastic, too? Which of all the samples has the most fuzz? What color is it?*

If fuzz is not yet apparent, put the bags back for a few more days. Then take them out once more. *Is the fuzz more noticeable now? Do you know of any nonliving things that increase like this in a few days without anything being added to them?*

Without removing the samples from the bags, let the class examine the fuzz with hand magnifiers or microscopes. *What does it look like when magnified? What changes have occurred in the food itself?* Put the bags away for another week and see what happens.

The fuzz is a mold that grows from microscopic *spores* that drift in the air. Some land on foods and, if conditions are right, they grow into fuzzy or powdery plants. Most are harmless, but it is probably best to discard the bags, unopened, when you are finished with them.

SANITATION SQUAD

When plants or animals die, the materials in them must be made available to living plants and animals for use. Otherwise, in time there would be a shortage of the substances needed for life. Decay is one process by which once-living plants and animals are recycled—becoming food for living ones, either by being eaten directly or by being returned to the soil, water, and air.

Pupils should have an opportunity to observe what happens to an animal after it dies and to see some of the creatures that help speed this important process of decay. To provide this opportunity, get a freshly killed mouse, bird, or other small animal (as may sometimes be found along the highway) or ½ pound of the cheapest raw meat. Go out with your class and place it on the ground at some distance from the school building, where it will not likely be disturbed by people. Cover it with a piece of ½-inch mesh hardware cloth, staking down the corners firmly so that dogs and cats will not dig up and carry off the carcass.

Take your class to visit the carcass each day for two weeks and then weekly for a month or two. Let the pupils observe and record as best they can its changes and the visitors to it.

When does the carcass first become smelly? How soon do flies visit it? How do you suppose they found it? (Let some of the class test their suggestions by experimenting.) *What parts show the first signs of decay? What parts do you think will be the last ones to decay?*

At each visit, let the pupils loosen the stakes and use a stick to turn over the carcass. *What kinds of creatures are at work beneath it? Do they seem upset at being uncovered? Where might they have come from? How may they have located the food?*

These animals are part of an efficient sanitation squad, most of whom work unseen—inside or under a carcass, or at night when they themselves are less likely to become food for bigger animals.

BURYING BEETLE

MAGGOT

Chapter 5

Water and Other Liquids

Water is important for pupils to study, not only because it is essential for life, but because it is so abundant, covering three times as much of the earth's surface as land does. It is the most common substance in living plant and animal tissue; without it, most other chemicals needed for life could not get into or out of cells. Its buoyancy makes it possible to move people and cargoes by ship—without which early exploration and commerce would have been impossible. Hydroelectric stations are and will continue to be a major source of electrical energy. Water is also important for sanitation and recreation.

Early experiences with liquids should include those dealing with buoyancy, change of state, solvency, and surface films. Children should investigate under what conditions objects sink or float, both in water and other liquids. They should watch ice melt, and water freeze, evaporate, and condense. They should also observe other substances that change state easily, such as candle wax, oleomargarine, and butter. And they should observe the wide range of chemicals that dissolve in water.

In their investigations, pupils should become aware of the ease with which certain liquids evaporate, and of the danger of fumes such as those from cleaning fluids and gasoline. They should experience the pleasure of playing in and around water, but they should also be made aware of the hazards and be taught to play safely.

Pupils must learn of society's increasing demands upon and for fresh water and develop a healthy concern for how it is used. The amount of water on earth does not change; it is only recycled. What is used by pupils today is part of the same water used by Columbus, Cleopatra, and even dinosaurs!

Finally, pupils should come to appreciate the vastness of the sea and the method by which fresh water is transported from the sea to land areas. They should know the potential of the sea—for all practical purposes, the last frontier on earth—for additional fresh water, for food, and for chemicals. They also should understand the importance of the sea for transporation and recreation, as well as for its effect on weather.

SOME IMPORTANT OBJECTIVES

Attitudes and Appreciations to Be Encouraged

Without water, there could be no life as we know it, since water is the major constituent of cells and tissues.

Clean water is not unlimited; hence, no person has the right to pollute or to waste it.

All the water used by a person or a community must eventually be used by others; so every user should be considerate of downstream neighbors and of future generations.

As a burgeoning population demands increasing amounts of fresh water, people must look to the sea for extracting additional amounts.

Although fossil fuels and nuclear energy are important, falling water from dammed-up rivers and tides is also a very important source of electrical energy. At the same time, most of the earth's hydroelectric potential is still untapped.

Some of nature's most beautiful and awesome displays—rainbows, seascapes, waterfalls, clouds, and thunderstorms—depend upon water or upon the energy released during the recycling of water.

Evaporation from the sea and subsequent precipitation over land is the means of replenishment of fresh water for drinking, sanitation, recreation, and irrigation, and of falling water for hydroelectric power plants.

Much of the sculpturing of the earth's surface has been caused by water in either its liquid or solid state.

Skills and Habits to Be Developed

Keeping streams, lakes, and oceans free of litter and pollution

Taking care not to waste tap water, even though the supply may seem limitless

Pouring liquids without spilling or splashing

Selecting the most absorbent material available and using it effectively to remove spilled liquids

Hastening the evaporation of a liquid by heating it, increasing its surface area, or speeding the movement of air across its surface

Using to advantage the buoyant force of liquids to float objects or to decrease their apparent weight

Covering cold objects such as water pipes and iced drinks to minimize unwanted condensation on them

Remembering to give water to pets, household plants, farm animals, and even wild birds as they need it

Using correctly terms such as *fluid, displacement, buoyancy, change of state, vapor, condense, cohere, surface film, immiscible, pollution, sewage,* and *conservation*

Facts and Principles to Be Taught

Water is rarely pure; it nearly always has other substances, even air, dissolved in it or mixed with it.

The tendency of some liquids to evaporate is so great that they exert considerable pressure in doing so.

Most substances are visible only in their solid or liquid state, not as vapors.

When a liquid evaporates, it leaves behind most of its impurities.

Different liquids vary in their tendency to evaporate; under the same conditions, some evaporate faster than water, and some less rapidly.

Many substances that are solid at room temperature become liquid and perhaps even gaseous at higher temperatures, while some that are gaseous at room temperature become liquid with sufficient cooling. Water is one of the relatively few substances that are liquid at ordinary temperatures.

Water and other liquids tend to cohere and thus to have a surface film that can support relatively heavy objects.

A liquid tends to buoy an object immersed in it with a force equal to the weight of the liquid that the object displaces.

When objects of the same weight are compared, those with larger volumes have greater buoyancy.

Large animals such as whales, and flimsy ones such as jellyfish, can live only in water, which buoys them.

LIQUID LIFT

Is it easier to float in fresh water or in salt water?
If pupils do not know, let them investigate!

First have them collect several short pencils,
about 8 to 10 centimeters long, and some olive
bottles, toothbrush tubes, or similar containers.
Supplement these containers with test tubes, if
necessary.

Next, divide the class into small groups. See
that each group has a pencil, an olive bottle or
similar container, and a paper cup. Then let them
proceed as follows.

1 Fill the container with enough water to float
the pencil, eraser end down. Hold the container
upright, and tap it to be sure that the pencil floats
freely. If it does not float in an upright position,
stick a thumbtack in its eraser.

2 Note where the water level comes on the label
of the pencil. *Does this change when you add more
water to the container? Does it change if you float
the pencil in a different container of water?*

3 Fill the paper cup with water, add a teaspoon
of table salt, and stir until the salt dissolves. Pour
this salt water into the olive bottle or other container, and float the pencil in it. *Does the pencil
float higher or lower than it did in fresh water?
Where does the level of the salt water come on it?*

4 Dissolve a second teaspoon of salt in the salt
water. *What does this do to its* buoyancy? In this
way, one can also test the buoyancy of sea water.
(See "Super Solution," page 73.)

*What do you suppose it would be like to swim
in very salty water, such as Great Salt Lake or the
Dead Sea? What must happen to ships as they sail
from the ocean into a body of fresh water, such as
the St. Lawrence River or the Panama Canal?*

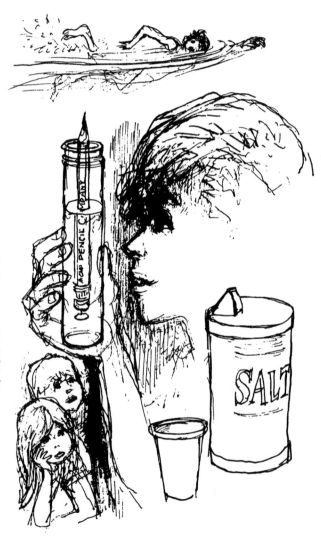

DIVING DROPPER

This amazing device fascinates children and helps them to be observant. It is a simple form of a toy that is often in the shape of a little devil or diver.

To make one, fill a medicine dropper part way with water. Change the amount of water in it until it *barely* floats in a tall glass of water. Only the very end of the rubber bulb should stick up above the surface of the water.

Then, without squeezing the bulb, transfer the dropper to a tall bottle filled to the brim with water. Place the palm of your hand tightly over the mouth of the bottle and press down on the water. The dropper will sink. Let go, and it will rise.

With practice, you—and your pupils—can make the dropper dive to any depth you wish, stay there, or rise. Children, instead of pressing directly on the water, may find it easier to push on a piece of rubber cut from a balloon. Stretch it over the mouth of the bottle, and hold it on tightly with rubber bands.

How the device works can be explained in two ways, really not different:

1 The pressure of one's hand causes a little more water to enter the dropper. This can easily be seen. *How does the added water affect the weight of the dropper? And so, what does the dropper do? Then, when the pressure is released, what happens to the added water? To the weight of the dropper?*

2 The air inside the dropper is really a bubble, and this tends to float because it is buoyed up by the water. The larger this bubble is, the more it is buoyed up. At first it is large enough—and buoyed up enough—to keep the dropper afloat. *But what happens to the size of the bubble when the pressure of the water on it changes? And how does this affect how much it is buoyed up?*

The same principle is used in submarines. In order to submerge, these take in water. And, in order to rise to the surface, they force it out again with compressed air.

HOLEY SCOW

A lightweight aluminum pan makes a good barge, or scow, for doing experiments. With it, pupils can investigate questions like these:

1 How heavy a load can such a scow carry? How many pebbles, nails, or paper clips are needed to sink it?

2 Will it float even if you make small holes in it with a needle? If so, will it still be able to carry a load?

3 With more numerous needle holes, will it sink? With larger holes, made with a sharp pencil?

4 When the scow is out of water and dry, what happens to drops of water dripped on the holes? What if this is done while the scow is floating?

5 While the scow is floating, are the water surfaces inside the holes flat or curved? If curved, how do they curve?

Water, like other liquids, acts as though it has a "skin" on its surface. This is the *surface film,* caused by the minute particles of water clinging together.

Children can also observe the surface film of water and its effects if they:

1 Sprinkle drops of water on a flat sheet of waxed paper.

2 Turn off a faucet gradually, so that the stream of water becomes thinner and thinner and then separates into drops.

3 Dip a strainer or wire screen into water; then lift it out.

4 Show how a person who falls in the water can keep afloat by trapping air inside a wet shirt.

5 Lay a piece of aluminum foil on water; then push it under.

6 Sprinkle dry sand on water to see if some grains "float."

7 Rub a thin needle with dry fingers; then "float" it on water by lowering it onto the surface with a bent wire.

8 Watch water striders "skating" on water, making dents in its surface with their legs.

ORANGE-PEEL MIST

A fundamental idea in science, based on observation coupled with imagination, is that substances consist of extremely small particles. This is the basis of our concepts of atoms and molecules—tremendously important in today's world.

A brief experience with a common liquid can make this basic idea of the particulate nature of matter more real to children. (This can also be done with solids and gases, as in "Lead Pencils," page 69, and "Vapor Push," page 63.)

Simply have each child take a small piece of orange peel, fold it over, and squeeze it. Tiny droplets of liquid will squirt out, forming a fine mist. This mist shows up best in bright light—perhaps in the beam of a projector in a darkened room. The children then can see that the droplets are indeed extremely small and numerous.

On the basis of what you observe, can you imagine still smaller droplets of this liquid? Can you imagine some so small that you could not see them, even with a microscope? Then, even if you could not see them, might you still smell them?

THICK AND THIN

Pupils probably know that some liquids are "thicker" than others. But do they know why?

Have teams use pairs of olive bottles or similar containers—perhaps test tubes—and pans to protect the tables or desks. Let each team fill one container with water and the other with salad oil, mineral oil, syrup, or other "thick" liquid.

Then, taking turns, pupils should drop identical paper clips, staples, or pins into the liquids. They may also try sand grains and tiny pebbles. (See "Sorted Stones," page 87.) *What happens? As an object sinks, what must it do to the minute particles of liquid? In which liquid are these pushed apart more easily?*

Bubbles, to rise through liquids, likewise must push apart the minute particles. *How easily can they do this in various liquids?* Pupils can make small bubbles in the liquids by vigorous shaking.

Several matched containers of different liquids, with a similar pebble or bubble in each, may be capped tightly. Then, for a fair test, one need only invert them.

VAPOR PUSH

In ''One-Gallon Cloud'' (page 37), pupils can learn one way that water comes *out* of the air. But water and some other liquids go *into* the air, too, as shown by the pressure they create when they evaporate.

Give each group of about six pupils a glass, a widemouthed gallon jar, some stout rubber bands, and a large, round balloon such as can be purchased in most supermarkets. Have one pupil in each group cut off the neck half of the balloon to leave a sheet of rubber. Then have pupils practice stretching this rubber sheet across the mouth of the jar so that it is taut. When they have demonstrated that they can do this quickly, ask them to proceed as follows.

Pour about 2 tablespoons of water into the jar and quickly stretch the rubber sheet across its mouth. Fasten it firmly with the rubber bands. Swish the water around in the jar a few times. Now sight across the top of the jar. *Is the stretched rubber perfectly flat?* Check it carefully at the end of each minute for 5 minutes. *Is it flat at the end 1 minute? At the end of 5 minutes? What might account for this? How does the appearance of the stretched rubber compare with that on an identical, but waterless, jar set up as a control?*

Some pupils may suggest that the air in the jar has warmed and expanded, pushing up the rubber. Let each group empty and dry its jar, put a thermometer inside the jar, and repeat the activity. *Does the stretched rubber bulge as before? Is this due to warming of the air?*

Now have pupils dry their jars thoroughly and replace the water with an equal amount of rubbing alcohol. *How does its effect on the rubber compare with that of water? How would rubber sheets stretched across two jars—one with water and one with alcohol—compare if left overnight?*

As water and other liquids evaporate, they take up more space. Without any change in temperature, air with water vapor in it takes up more space than the same air before water vapor is added. Suppose that the door and windows of a classroom are closed and the chalkboard is moistened. *What would happen in the space under the classroom door as a result of the evaporation from the board?*

BENEATH THE SUDS

Most children know that soaps and detergents help
to get rid of grease and oil on hands and dishes,
but probably few know how they do it. With two
immiscible (unmixable) liquids such as oil and
water, and a small amount of detergent, pupils can
find out for themselves what happens beneath the
suds.

See that each pupil has a small, screw-cap transparent jar, a toothpick, and a paper towel for spills.
At one side of the room place several small containers of cooking oil, each with a dropper. Also
set out a few containers with small amounts of
detergent. Ask the pupils to fill their jars two-thirds
full of water and add 10 to 20 drops of oil. Then
have them take their jars back to their seats to
investigate as follows.

*Does the oil float or sink, or does some float
and some sink? If some is pushed under the water
and released, what happens to it? If the oil is
stirred gently, does it mix with water?* Cover the
jar, shake the oil and water, and let it stand. *What
happens to the droplets? Can you tease these back
together again?*

Now dip the toothpick into some detergent, and
then touch it to the center of the water in your jar,
watching carefully as you do so. *What happens at
the oil-water surface?*

Cover the jar. Then shake the oil and water a
few times. Hold the jar up to the light and observe
the contents carefully, using a magnifier if one is
available. *Are there more or fewer droplets of oil
than before? Are they larger or smaller?* Add a
tiny bit more detergent, shake the mixture, and
observe it once more. *What change can you see
in the number and size of the oil droplets?* Try
other immiscible substances, such as motor oil,
kerosene, and melted butter or oleo.

Detergents and soaps help to break oil and
grease into tiny droplets. When the droplets are
small enough, they do not stick to hands and
dishes, but wash away easily in water.

LADDER FOR LIQUIDS

In "Holey Scow" (page 60), water was observed to cling to itself, or *cohere*. However, many liquids, including water, also stick to other materials, or *adhere*. Depending upon their abilities both to cohere and to adhere, liquids may do interesting things such as climb inside slender tubes, creep up or down the sides of containers, and soak into porous materials. Besides being useful in lamp and candle wicks, sponges, and soil, this movement of liquids, called *capillary action,* is an intriguing phenomenon for children to investigate.

Let each pupil in a group cut a 25-centimeter strip from different materials, such as cloth, paper towel, cardboard, and string, and tape one end of each to a ruler resting across two books standing on end. Then have each pupil fill a paper cup with water to which some food coloring has been added for visibility. At a signal, let pupils put their cups under the strips and observe how the water climbs. *In which strip does the water rise fastest? In which has it risen farthest after 5 minutes? After ½ hour? What differences can be observed if cooking oil is used in place of water? If rubbing alcohol is used?*

Now let each group compare, as follows, the rate at which water moves up, and then *down,* liquid "ladders." Fill a cup with colored water and set it on another cup, which has been inverted. Cut several equal-size strips, as before, long enough for one end to dip into the full cup and the other end to dangle into an empty cup below. Cut a point on the dangling end. Put the strips in place and watch what happens.

Through which strips does the water soak to the point? Does the water stop moving when it has soaked to the point? At the end of ½ hour, what has happened in the empty cups? The full cups? What will happen if they are left overnight? What explanation can be offered for what is observed?

Suppose a person using a pail of liquid and a cloth left them on the floor overnight. *What might happen to the liquid in the pail if one end of the cloth touched it and the other end dangled over the side?*

Powders and Solutions

Countless materials that we take for granted, including soft drinks, detergents, and medicines, are products of chemistry. Many common processes, too, such as burning, rusting, and cooking, involve chemical action. In such processes, substances act chemically with other substances, or else they decompose, and new products are formed; *chemical changes* are involved. In other common processes, however, no chemical action takes place, and no new substances are produced; only *physical changes* occur.

With some understanding of chemical and physical changes, a person can better appreciate the materials and processes encountered in daily living. Such understanding may start early in life—perhaps through making lemonade, performing magic tricks, or playing with a chemistry set. In school, however, pupils are too often presented with theoretical concepts, like those of atoms and molecules, without adequate opportunity for first-hand experiences, such as with household powders and solutions. It would be better for them if the theory were introduced after, or along with, such experiences. Theory must relate to what is observed; otherwise it means little to children.

Basic experiences with powders and solutions do not require test tubes, beakers, and Bunsen burners; ordinary containers and sources of heat usually suffice. Neither are formulas and equations necessary—although the chemical names of some common substances may be introduced.

Safety measures are important, of course. However, having to take precautions should not bar pupils from experiencing chemical and physical changes—no more than having to obey traffic lights should keep them from crossing streets.

The thrill of watching changes take place—with predictable results—makes working with powders and solutions fun. This is true for the teacher and pupils alike. Besides, with activities that require only everyday things, both will be encouraged to extend their investigations at home.

SOME IMPORTANT OBJECTIVES

Attitudes and Appreciations to Be Encouraged

All substances—even ordinary things such as air, water, salt, paper, and foods—consist of chemicals.

Most commonplace chemicals are not especially dangerous; some, however, are poisonous, caustic, flammable, or explosive, and so must be stored and used with great care.

If two or more samples of chemicals have exactly the same characteristics, and act the same in every way when tested, they are assumed to be the same substance.

Some changes in substances, called *chemical changes,* produce new substances, with new characteristics; other changes, called *physical changes,* do not produce any new substances.

Chemical changes are taking place all around us—for example, in cooking, burning, and rusting—and are not confined to test tubes, beakers, and flasks, or to laboratories.

In general, chemical changes are predictable, allowing chemists to anticipate what will happen and what the results will be.

The results of chemical changes may be harmful or helpful, depending on how they affect human beings and other living things.

During chemical changes, some substances are formed and others are destroyed, but there is neither creation nor annihilation of the "stuff" or matter of which the substances consist.

To chemists we owe many new products, including modern plastics, synthetic fibers, and wonder drugs—with better things yet to come.

Because of chemistry, people are faced with a dilemma: new materials that decay, disintegrate, or corrode more slowly result in longer-lasting products, but also in longer-lasting refuse and litter.

The indiscriminate use of chemicals and the improper disposal of chemical wastes cause very serious pollution of air, water, land, and living things, and result in human illness, economic loss, and environmental ugliness.

Skills and Habits to Be Developed

Recognizing and heeding the warning labels on containers of poisonous, caustic, flammable, or otherwise dangerous chemicals

Following directions carefully.

Using dangerous chemicals only under the supervision of a responsible adult, with great care and with water at hand for washing or putting out fire in an emergency

Experimenting very cautiously with unfamiliar substances, using only small amounts

Labeling containers and samples of chemicals accurately and legibly

Taking samples from containers without contaminating the supplies

Removing all traces of liquid or solid substances from one's hands, from glassware and other utensils, and from the workplace

Measuring out volumes of liquids accurately, both by counting drops and by using calibrated containers

Weighing out samples of solids or liquids accurately, with both homemade and commercially made balances

Determining the volumes of gases indirectly by means of liquids that displace, or are displaced by, the gases

Pouring and mixing liquids and making and diluting solutions—without spilling, splashing, or contaminating them

Distinguishing between undissolved and insoluble material, and detecting very small amounts of dissolved solids in water

Heating substances safely by means of an electric hot plate or stove, candle, or alcohol flame

Identifying common household powders from their characteristics, such as solubility in water or action with vinegar, tincture of iodine, or red cabbage juice

Testing substances for acidity or alkalinity with indicators such as red cabbage juice or grape juice

Carrying out simple research projects in chemistry—concerning, for example, air and the rusting of iron, sea water and the sudsing of soap, or lemon juice and the charring of paper by heat

Timing and graphing the life of a candle flame when covered by jars of different sizes, and predicting its life expectancy beforehand from the volume of a jar

Providing sufficient air (more specifically, oxygen) for a fire that is wanted, and cutting it off from one that is not wanted

Reading, with understanding, such terms as *acid, baking soda* (or *bicarbonate of soda*), *chemical change, dilution, extinguish, indicator, liter, oxygen, poison, solution*

Facts and Principles to Be Taught

Evidence indicates that chalk, charcoal, and copper—and, in fact, all substances—consist of very tiny particles.

Chemical tests, such as whether there is bubbling with vinegar or a change in the color of tincture of iodine, can be used to distinguish among substances that otherwise seem alike.

Air, because of the oxygen it contains, is necessary for both ordinary burning and rusting, as well as for living.

Oxygen makes up about one-fifth of the volume of ordinary air; the remainder is nearly all nitrogen.

Fires can be extinguished by surrounding the burning material with something that keeps air away but that, itself, does not burn.

Carbon dioxide is a common and easily made gas, much used for extinguishing fires, raising dough, and making soda pop fizz.

Even perfectly clear water usually contains dissolved substances, and when it evaporates, any dissolved solids in it are left behind.

Sea water is a solution of various chemicals, mostly common salt, which have been washed into the ocean from the land and air.

When dissolved substances come out of solution as solids they usually appear in the form of crystals, and these tend to have characteristic shapes.

Solids are generally more soluble in warm water than in cold; gases, however, are more soluble in cold water than in warm, and their solubility increases with pressure.

Some chemical changes, as in fires and explosions, are rapid, but others, like decay and rusting, are much slower.

How rapidly chemical changes take place depends upon both the substances involved and the temperature; some substances change extremely slowly or not at all, even at high temperatures.

Water is often necessary for chemical changes to occur—for example, chemical powders usually will not act on each other until water is added.

Acid and alkaline substances have different effects, as on the color of indicators like red cabbage juice or litmus.

SECRET INK

To children, sending secret messages is very exciting. One way to do this is with invisible ink.

This may be a common colorless or nearly colorless solution, such as lemon juice. It can be used to write a message on ordinary paper with a clean pen point or toothpick.

Then, to make the message visible, the recipient heats the paper until it is quite hot. This can be done safely in a disposable aluminum pan on a hot plate or stove.

Can you discover an invisible ink of your own? Let pupils test various common liquids, such as fruit and vegetable juices, vinegar, soft drinks, and solutions of sugar, salt, and baking powder.

Why does the writing appear when the paper is heated? We authors do not know. Perhaps something dissolved in the ink decomposes, and chars sooner than the paper. Or something in the ink may absorb radiant energy better and heat up faster than the paper, causing the paper to char. *Does a bit of sugar char when heated in a pan? A pinch of salt or baking soda?*

LEAD PENCILS

When a fishing sinker or other piece of lead metal is rubbed on paper, it leaves a mark. Pupils can write with it. They may even hammer lead sinkers into long "pencils"!

Many years ago, lead metal was used for drawing and writing. However, the "lead" in modern pencils is *graphite,* a form of carbon.

Lead and graphite make marks on paper because tiny particles rub off as a fine powder. *Are you able to see these particles with a magnifier? Might there be particles even smaller than any you can see?*

Children can rub other substances on paper and observe the results. They may try aluminum foil, copper wire, iron nails, and brass screws—even old silver spoons and gold rings. They also may try chalk, charcoal, wax, rust, and brick.

Do tiny particles rub off all these substances equally easily? What happens when the substances are bent, or hit with a hammer? Do their particles separate? Or do they move in relation to one another, yet still hold together?

CHEMICAL CLUES

Chemists are often asked to identify unknown substances. To do this they look for clues—somewhat like detectives. Doing detective work of this sort with common powders is fun for pupils. It also gives them some idea of how chemists work.

Baking soda, corn starch, and cream of tartar are white powders found in many homes. *Can you tell them apart just by their appearance? If not, how might you identify them?* Small teams of pupils can make tests and find clues as to how these powders differ. Later, from these clues, they can detect the powders.

First have a pupil from each team put ½ teaspoon of baking soda into each of three glasses or cups half filled with water, and stir until dissolved. Then ask a second pupil to add a little vinegar to one glass or cup. *What happens?* Have the pupils keep a record.

Next, let a third pupil from each team add a few drops of tincture of iodine to the next glass or cup. *What change takes place?* Then ask a fourth pupil to add a pinch of baking soda to the last container. *What, if anything, happens?* Again, the pupils should keep records.

Now let the teams repeat these tests with corn starch in place of baking soda. *Does it dissolve as completely as baking soda? Do bubbles form when vinegar is added to it? Does it change the color of tincture of iodine? Does anything happen when baking soda is added to it?*

Then the teams should apply the same tests to cream of tartar. As before, the pupils should keep records of their observations.

After this, let the teams test "unknown" samples containing one or more of the three powders. (Make these up beforehand, and record their contents.) *Which powder, or powders, does each sample contain?*

Later, pupils may also test small samples of other nondangerous household substances, such as baking powder, pancake mix, tooth powder, talcum powder, and antacid tablets. *Which of the three powders* may *be present in each? Which of them* cannot *be present?*

FLAME LIFE

A candle needs air to burn. When a lighted candle is covered with a jar, the flame soon "dies."

What effect, then, do you think the size of the jar will have on the life of the flame? A class can check this with an assortment of jars—quart or liter, and smaller.

Let the pupils work in teams. Each team should stand a birthday-cake candle in a wad of aluminum foil on a smooth desk top or table top. Appoint a "Candle-keeper" for each candle, to light it and to put it out if necessary. Show how to do these things properly. Also have a can of water near each candle.

Then appoint a "Timekeeper" for the class. He or she is to call "Ready! Set! Go!" and at "Go!" to start counting seconds aloud. (See "Swinging Second-Timer," page 19.)

Also at "Go!" someone in each team should set a jar over the lighted candle, and start a count of how many seconds the flame continues to burn. Then each team member should repeat this—after fanning fresh air into the jar—and make a record of how long the flame "lives."

Next, each team should find the volume of air in its jar by pouring in water from a small paper cup or a measuring cup. This volume—in cups, fluid ounces, or milliliters—should be checked and recorded.

Now the relation of air volume to flame life may be shown by a graph. For this, rule a large grid. Number the horizontal lines in units of volume and the vertical lines in seconds, by tens. Then have the teams make X's on this grid to show the air volume in each jar and the average flame life of a candle beneath the jar. Draw a smooth line through these X's so that about as many of those it misses are on one side as on the other.

According to this graph, what will be the life expectancy of a candle flame beneath a 2-quart jar or a 2-liter jar? Beneath a 1-gallon jar? To help pupils make predictions, extend the line on the graph. *Do actual tests show your predictions to be correct? If not, why not—what else might affect flame life? How could you check to find out?*

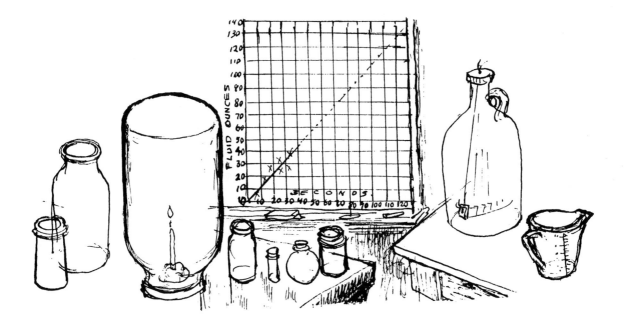

FLAME KILLER

Carbon dioxide is an interesting, safe, and useful gas—easily made from common substances. It is often used to extinguish fires. Pupils should know why, and also how.

To prepare some carbon dioxide, pour ¼ cup of vinegar into a pint jar. Add 1 teaspoon of baking soda and quickly cover the jar with a flat piece of cardboard. **Do not screw a lid on the jar!**

Next, stand a candle in a pan and light it. Uncover the jar and immediately pour its contents—*but not the liquid or solid contents*—on the flame. Hold the jar so that its mouth is a little higher than the flame and a short distance away. *What happens? Can anyone see what causes this?*

Let pupils try this under supervision. Each one should have a turn. However, have only one or a few lighted candles in the room—each guarded by a responsible Candle-keeper who has a can of water at hand, as a precaution.

Vinegar (a weak solution of acetic acid) and baking soda (sodium bicarbonate) act together chemically and produce carbon dioxide. This gas "kills" flames by keeping air away. To show why it can do this:

1 Suspend a long, thin stick by a single thread tied to its middle. Open two paper bags, and hang them from the ends of the stick so that they balance. Then prepare a few jarfuls of carbon dioxide and pour them into one bag. *What happens? What does this show?*

2 Make a long handle of wire for a candle. Light the candle and lower it into a jar of carbon dioxide. *What happens? Why?* Then pour the gas out of the jar and try this again.

3 Put some baking soda into a balloon and add some vinegar. Hold the mouth of the balloon shut. *What happens? Why? What happens when you aim the mouth of the balloon downward and let it open slightly?*

Most fire extinguishers shoot out carbon dioxide, often mixed with water or foam. It is good to have a custodian show the class how to use them, on a small fire outdoors.

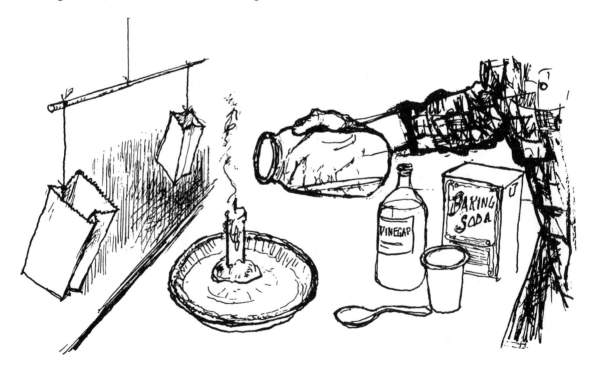

SUPER SOLUTION

There is more of one liquid solution on earth than of any other. In fact, there is more of it than of *all* other liquid solutions put together!

Who can think what this solution is? If pupils are baffled, give them some clues, one at a time, such as: It has a taste you will never forget. Fish can live in it. Most of the earth is covered by it!

Although children read about the sea, many who live inland do not know what sea water is like. Let them find out. They will learn other things, too, at the same time.

The class may write to the same grade in some coastal village and request ½ liter or so of *clean* sea water—perhaps as a swap for fossils, minerals, or other interesting objects. It can be mailed in a *plastic* bottle with a tight screw cap, packed well with newspaper in a strong carton.

When the sea water arrives, teams of pupils may carry out various investigations with it, as follows:

1 Pour a little into a jar, set the jar in a pan of water, and boil the water for 20 minutes. *After it cools, what does the sterilized sample taste like?*

2 Put some in a glass dish, let it evaporate, and examine what is left with a magnifier. *What happens as more sea water is added from time to time and allowed to evaporate?* (See ''Reappearing Act,'' page 112.)

3 Try making soapsuds in sea water, and also in a solution of *pure* table salt (sodium chloride) in rain water. *Judging from what happens, is sodium chloride the only dissolved substance in sea water?*

4 Compare the rusting of nails in sea water and fresh water, and also test items of aluminum, brass, and copper in both. *Of what practical importance are your findings?*

5 Shake equal pinches of mud in equal amounts of sea water, rain water, and creek water, and compare how quickly the finest particles settle. *How might this explain what occurs when rivers carry mud into the sea?*

INVISIBLE SOLUTION

In "Super Solution," pupils are asked to think of the most abundant liquid solution on earth. However, there is even more of another solution, this one *not* a liquid. It is air, a solution of gases.

Pupils can easily obtain evidence that the solution we call *air* has at least two different gases in it. For this, ask them to bring olive jars, vials, toothbrush tubes, or other slender containers, and also jars or tumblers. And have them, at odd times, rub a large nail on a flat file to produce iron powder.

Next, ask each pupil to wet the inside of a slender container and add a large pinch of iron powder so that it sticks to the glass. He or she should then set the container, mouth down, in some water in a jar or tumbler.

What keeps the water from rising higher in the containers? How could you tell if some of the air inside a container were removed?

Now let the class observe the containers for several days. *What happens to the water levels? What does this suggest? Does the water, in time, fill the containers, or does it rise only so far?*

Have the class observe the iron powder, too. *Is there any sign that it is being changed to a new substance? If so, is all of it changed, or is some left unchanged? How could a magnet be used to check this?*

After a few days, let the pupils estimate, or measure, what fraction of the volume of each container has filled with water, and record this. *Is there general agreement? What seems to have caused the change in water levels? Might any other factors have affected them?* (See "Bottle Barometer," page 36.) *What if no iron had been used?* Pupils should check this by experiment.

In rusting, the iron combines with part of the air and removes it; this is the gas called *oxygen*. The gas that is left does not combine with the iron; this is nearly all *nitrogen*. *Based on your findings, about how much of air is oxygen? About how much is nitrogen?*

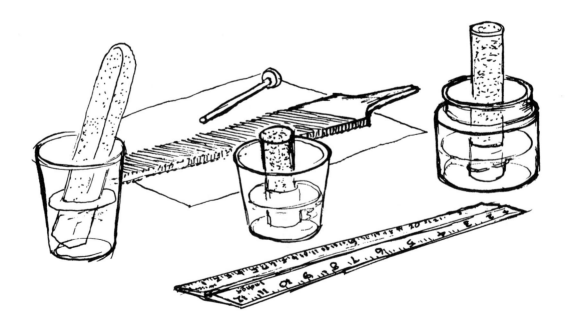

ONE IN A MILLION

When you rinse out a glass, what are you really doing? When is there no trace at all of what was in the glass? With some simple materials, pupils can learn that rinsing really is diluting—to one part in a million, or even less.

Using a medicine dropper, count out 99 drops of water in a clean paper cup. Then add 1 drop of food coloring or ink. Remove the rubber bulb from the medicine dropper and let water run through and over the glass tube to wash away all traces of coloring. Replace the bulb. Then stir the mixture in the cup. The cup now has 100 drops of liquid, of which 1 is coloring. This ratio of 1:100 is the *dilution* of the coloring.

Put a drop of this mixture into a second cup, add 99 drops of water, and stir it. Now the dilution is 1 drop of 1:100 *in* 100, or 1 in 10,000 (1:10,000).

Put a drop of this mixture into a third cup, add 99 drops of water, and stir once more. The dilution in this third cup is 1 drop of 1:10,000 in 100, or 1:1,000,000! In a fourth cup put 100 drops of water, and compare its appearance with that of the third cup. *Can you see any difference between them?* You know there is a tiny, tiny bit of coloring in the third cup because you put it there. *How long do you think you could keep diluting this mixture before there was absolutely* no *coloring left in it?*

Pour out the mixture that is in the third cup. *Do the drops sticking to the bottom or sides of the cup contain any coloring at all?* Imagine how dilute the mixture would be if you filled the cup to the top with water again before emptying it!

Let some blindfolded pupils take a taste test of dilute solutions of lemon juice and of honey. Prepare dilutions of 1:100, 1:500, 1:1,000, and so on. Put a medicine dropper in each cup for dropping the solutions on the pupils' tongues. Then find out how dilute a solution the pupils can identify correctly three out of four tries. *Are pupils more sensitive to dilute solutions of something sour or something sweet? Even when they cannot taste the lemon juice or honey, is there still some in the cup? How do you know?*

Some animals such as ants can detect solutions far more dilute than those human beings can taste. To find out how dilute, prepare a dilution of honey that is 1:1,000,000. Put this sample, and an equal sample of water, in identical bottle caps that have been thoroughly cleaned and then rinsed in running water. Set them near a place where there are ants.

The next day, examine the caps to see if there is a noticeable difference in the number of ants visiting each one. If not, replace the cap having the dilute solution with a fresh one that is 10 times as strong (only 1:100,000). Each day, examine the caps, replacing the dilute solution with a stronger one, until you can observe a difference in the number of ants visiting each cap.

What is the weakest (most dilute) solution that seems to make a difference to the ants? If everything else is the same except for the dilution of the honey, what might be causing what you observe? From your observation, how dilute a solution of honey can ants detect?

BUBBLES FROM NOWHERE

Most pupils know that there is gas in soda pop; they have seen the bubbles and felt the fizz. However, they may not know that water often contains gas, too, which comes out under certain conditions. Pupils working alone or in small groups, each with two drinking glasses, a test tube, a bottle of clear pop (such as ginger ale or 7-Up), a hand magnifier, and a thermometer, can investigate gas leaving a solution, as follows.

Fill both glasses with cold water and record its temperature. Label each glass with this temperature and the names of the pupils in the group. Leave one glass in the room at a place where it will not be disturbed. Place the other in a refrigerator. After an hour, examine the glass in the room. *Can you see any bubbles in it?* Take the temperature of the water. *How does it compare with the original?*

Now examine the glass in the refrigerator. *Are there bubbles in it? How do they compare with those in the other glass? How does the temperature of this water compare with the other?*

Let the pupils propose explanations for what they see. On this basis, let them try predicting what will happen if the glasses are filled at the start with warm water instead of cold. Then let them try it. *How do the observed results compare with the predicted results?* Let them try, also, glasses of cold and warm water placed in sunlight and in shade. *Does the light make a difference, or does sunlight merely warm the water and thus cause bubbles to form? How can you test your ideas?*

Have each group fill its test tube completely with cold tap water. Then have one pupil put a fingertip in the mouth of the test tube and pry it upward to reduce the pressure inside. *Do any bubbles form and rise?* Let the water stand until it is about room temperature. *Do more bubbles form?*

Tap water, like soda pop, is under pressure until it leaves its container (the pipe) through the faucet. Some of the air dissolved in it comes out—partly, perhaps, because the pressure is reduced, but mostly because the water warms. Warm water cannot hold as much dissolved gas as cold water can. Bubbles can be seen forming in water heating on a stove, before the water has begun to boil. The bubbles may be air, or they may be water vapor (steam). *Can anyone think of a way to collect some of the bubbles and then test them to find out?* When a group has an idea, let that group try it (with supervision) for the rest of the class.

Soda pop is a solution of carbon dioxide in water, but since it is dissolved under pressure, much of it escapes when the pressure is reduced. Like air bubbles in water, carbon dioxide bubbles in soda pop seem to come from nowhere as the gas leaves its solution.

Let the pupils, working in small groups, investigate bubbles in pop as follows: Carefully remove the cap from a bottle of clear pop at room temperature. *What can you see happening in the pop?* Let someone put his or her mouth tightly over the opening of the upright bottle and force air into it as hard as possible. *What happens to the bubbling rate?* Then let the pupil try to suck air from the bottle. *How does this affect the bubbling rate?*

Pour some pop into a glass and look carefully through a magnifier at where the bubbles form. *Do they form at various places, or do they keep forming at particular spots? Do they suddenly appear as large bubbles, or do they grow from nothing to one big enough to break free and float to the top?*

What happens if you tap the glass with a pencil?

Some other things for pupils to try and to think about when investigating bubbles from nowhere:

1 Stick a finger in a glass of soda pop and compare the formation of bubbles on it with the formation on the glass itself.

2 Stick a small piece of adhesive tape to the inside bottom of a glass, pour pop into the glass, and see where bubbles form most readily.

3 Drop a few grains of salt into a glass of pop. *How do they affect bubble formation? Do sand grains do the same?*

4 Some fish, such as trout, normally live in cold water. *Why might they have difficulty living, and perhaps even die, if the water becomes warm?*

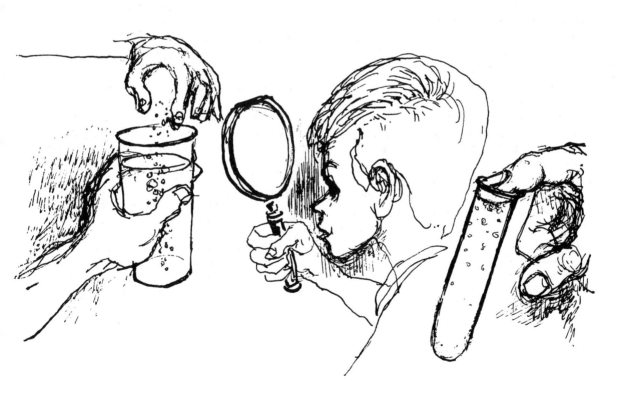

DOUGH RAISER

Baking powder is one of the more familiar chemical mixtures in the home. Pupils can make some of their own, and bake with it. Also, by experimenting, they can learn how—and why—it works.

First they need to weigh out the amounts, in either column below, of these powders:

cornstarch	25 units	22 units
baking soda	25 units	27 units
cream of tartar	50 units	45 units
tartaric acid		6 units

Food stores sell the first three. Some druggists have tartaric acid—not a dangerous acid, but sour!

The units are of weight or of mass. Standard units, such as grams, are fine. However, depending on how much baking powder is wanted, the weights of identical pins, paper clips, or nails— "pin weights," "paper-clip weights," or "nail weights"—are just as good. A simple balance (see "Three-Corner Balance," page 27) may be used.

Pupils should mix the powders well in a dry jar, and then use the mixture to bake with—perhaps at home. They also can compare their homemade baking powder with a store-bought brand. For instance, they may bake a batch of biscuits with each, *keeping everything else exactly the same.* Then a taste test will show which they prefer.

To show how the baking powder acts, drop a spoonful into a glass of soapy water. *What happens? What would such action do to dough?*

The bubbles are of carbon dioxide gas. Pupils can tell which of the ingredients produce this gas by testing them—singly and in various combinations. For example, they can mix cornstarch with water, add baking soda, and then cream of tartar. *Does any one ingredient, by itself, produce carbon dioxide, or does this take more than one? Must water be present, also? How many substances, at least, are needed? What are these substances?*

ACID OR ALKALINE?

Chemists, in their work, carry out many kinds of tests—some based on changes in color. One such test, using red cabbage juice, is great fun for pupils to try.

Boil a few red cabbage leaves in water and save the colored liquid. (For future use, some may be frozen into ice cubes.) Then, to make tests, dilute the liquid with water until the solution is fairly transparent. Pour samples into several glasses.

Keep one sample for comparison. Let a pupil add a little vinegar to a second, and stir. Have another pupil squeeze a lemon into a third. *In each case, what color does the solution become?* This color indicates that an *acid* substance was added.

Now let volunteers add baking soda, washing soda, and household ammonia to other samples. **Caution: Ammonia is caustic. Guard against letting it get on skin—and above all, in eyes. If any does, wash it off at once.**

What colors are produced? These colors indicate that *alkaline* substances were added. Purple indicates little or no alkalinity; blue indicates greater alkalinity; and green, still greater.

Because its color indicates acidity or alkalinity, red cabbage juice is called an *indicator.* To learn more about it and other indicators, pupils may:

1 Use red cabbage juice to test various everyday substances—fruit juices, soft drinks, soaps, detergents, shampoos, saliva, and even soils. *Which ones are acid? Alkaline? Neither?*

2 Add just enough ammonia to some of the juice to turn it green; then vinegar to make it red. *Can you change its color back to green? To red? Can you, at will, make the solution acid or alkaline?*

3 Try to discover some other indicators. *Do grape juice, blueberries, or colored flowers work? What is litmus paper, much used in laboratories? Can you find out where litmus comes from?*

Rocks and the Land

Children are fascinated by rocks, minerals, and fossils, and bring their finds to class at the slightest encouragement. They are intrigued, also, by mountains, canyons, and waterfalls. Their interest in these things is good and should be nourished.

After all, people depend on earth materials and landscape features for a great many of their needs. To cite a few examples, they require stone for buildings, soil for crops, petroleum for fuel, valleys for reservoirs, and beaches for recreation. How wisely they act in using and conserving earth materials, and in changing or preserving landscape features, is of major importance. Thus, it is essential that everyone have a basic understanding of geology—of rocks and the land.

A person's conception of the earth—its composition, the changes that occur on it and within it, and its long history—is based, in the last analysis, on firsthand experiences. Many of these experiences stem from commonplace counterparts of geological phenomena, including gullied lawns, broken sidewalks, and dried-up mud puddles. Others may come from stone walls, creek banks, and scenic overlooks. Interesting in themselves, such experiences also furnish a foundation for comprehending abstract and complex concepts about the earth. They provide a basis for understanding how canyons are formed, what causes earthquakes, why some rocks are in layers, and the like.

What is taught in the area of geology—perhaps to a greater degree than in other areas—should depend on what there is nearby. This differs, of course, from place to place—but there is *some* geology to be observed at first hand *wherever* one happens to be! Children should be encouraged to investigate rocks and the land with which they can have direct contact. This is far better than to emphasize mostly faraway geological phenomena, however spectacular, with which their experiences can only be vicarious.

SOME IMPORTANT OBJECTIVES

Attitudes and Appreciations to Be Encouraged

Rocks and the land are interesting, and they become even more interesting as one observes them and learns more about them.

Scientists find out about earth materials such as minerals, earth features such as caverns, and earth processes such as stream erosion, by making observations, doing experiments, and testing hypotheses.

In learning about rocks and the land, we must base our ideas on evidence, not on mere hearsay or snap judgments—and draw inferences and conclusions with caution.

Although scientists know a great deal about the earth—what it is made of, what happens to it, and what its long history has been—there is much more still to be learned.

Our knowledge of what the earth was like in the past—with long-gone seas, huge glaciers, and strange beasts—is only inference, based on incomplete information gathered from rocks, landscapes, and fossils.

In general, the processes such as erosion and deposition that acted on the earth in the past were similar to the processes that act on it today; thus, "the present is the key to the past."

For the most part, changes in rocks and the land have taken place slowly and gradually, although cataclysms have occurred and still do occur at times.

Even awesome features such as canyons, and catastrophic events such as earthquakes, are the result of natural processes; they can be explained without resorting to supernatural causes.

Some of the changes that people make and will increasingly be able to make in the land have, or will have, undesirable consequences.

For our present and future well-being, it is imperative that we conserve—use wisely and not waste—minerals, fossil fuels, soils, and other earth materials.

In the view of many persons, earth materials and the land, like air and water, belong to *all* the people and not just to some.

Skills and Habits to Be Developed

Breaking a pebble or other small piece of rock safely by enclosing it in a strong plastic bag, placing it on a large rock, and hitting it squarely with a hammer

Describing differences in the shape, smoothness, color, clarity, and composition of pebbles, sand grains, and other pieces of rock

Classifying pieces of rock according to size, as boulders, cobbles, pebbles, granules, or sand grains; and according to shape, as angular or rounded

Separating sand, silt, and clay by shaking a mixture of them in water and allowing the particles to settle

Observing that some rocks are being, or have been, broken, abraded, discolored, and decomposed by natural processes

Pointing out places where soil or loose rock material is being eroded and deposited, and estimating how rapidly this happens

Recognizing landforms such as canyons, deltas, fans, landslide areas, and former lake beds—both as miniature and as full-scale landscape features

Conducting simple but valid scientific research—for example, on soil erosion and factors affecting its rate

Noting layers of sediments and of sedimentary rocks, and giving evidence that these probably were deposited in water or by wind

Telling the order of deposition—the relative age—of undisturbed layers of sediments or sedimentary rocks

Finding fossils, and recognizing them as being the remains or other signs of animals or plants of the past

Measuring the volume of pore space in sand and gravel, and relating this to supplies of underground water

Using metric units of length—centimeters and millimeters; and of volume—liters and milliliters

Determining whether "solid" rocks have tiny pore spaces within them, and thus whether they might contain water, petroleum, or natural gas underground

Distinguishing between rocks which, like granite, consist of interlocking crystals of minerals, and those which, like sandstone, are composed of particles stuck together

Identifying some common kinds of rocks, such as sandstone, conglomerate, limestone, marble, and granite

Labeling rock and mineral specimens, fossils, and soil samples accurately and clearly

Using correctly such terms as *abrasion, bedrock, cavern, erosion, fossil, glacier, mineral, pore, silt, topsoil*

Facts and Principles to Be Taught

In nature, rocks are continually being broken, abraded, and decomposed—some more readily than others—forming smaller and smaller pieces.

Rain water, streams, and waves move soil and loose rock material, generally sorting the particles according to size and depositing them in layers.

The surface of the land is shaped by gravity, water, air, ice, and living things, but chiefly by moving water working together with gravity.

Although earth features such as mountains are continually being changed, the basic materials are not destroyed but are used over and over again.

Well-rounded pebbles are a sign of abrasion caused by water moving in streams or as waves, either at the present time or in the past.

The rapidity of soil erosion often depends on how steeply the land slopes, on how well it is covered by plants or dead plant material, and on how it is plowed.

Bits and pieces of rock come in a great variety of sizes, from huge boulders to minute particles of silt and clay.

Different sizes of rock particles settle at different rates in water—the smallest generally the slowest—and thus often form layers on the bottom.

Some rocks contain fossils—the remains and other evidences of plants and animals of the past.

The particles in sand and gravel, as well as those in many rocks such as sandstone, do not fit together tightly; underground, the spaces between them usually contain water but sometimes petroleum or natural gas.

Some kinds of rocks consist of bits and pieces of older rocks, or of plant or animal materials, which have become cemented or otherwise stuck together.

Many rocks, including granite, consist of interlocking crystals of minerals formed deep underground under great pressures and high temperatures, sometimes from molten material.

Different minerals often can be told apart on the basis of their color, shininess, and transparency, as well as from the way they break and their ability to scratch various other materials.

Some rocks, such as limestone and marble, are much used for crushed stone, building stone, and other purposes.

Limestone and marble are dissolved by water containing weak acids; in stronger acids, such as vinegar, they produce bubbles.

Glaciers are formed from snow that has gradually changed to ice, and they usually contain pieces and bits of rock.

Many earth processes, such as the solution of limestone, are slow, yet their effects are great; therefore, they must have been going on for a very long time.

PEBBLE JAR

In many areas one can find a fascinating variety of pebbles. Young children like to collect them. This may lead to making a collection in a jar, as an early step in learning about rocks.

Start with a tall, plain glass jar. Let each child find a colorful or otherwise interesting pebble, wash it well, and place it in the jar. If there is still room, ask the children whose pebbles do not show well to add others. Then, when the jar can hold no more pebbles, let someone fill it with water. *Why does the size of the pebbles seem to change?* (See "Big Finger," page 164.)

Stand the jar near a window, where there is good light. Have at least one magnifier handy. *Do the pebbles all look alike? Are they similar in shape and smoothness? What colors do they show? Do some have more than one color? Do any shine or sparkle? Does light shine* through *some?*

At this stage it probably is not important for pupils to learn the names of the different kinds of rocks. However, they may wish to make up appropriate names of their own, such as black-and-white-stone, glass-stone, ribbon-stone, yellow-stone, and sparkle-stone.

Suppose that the pebbles were broken into tiny pieces, about like sand. *How many different kinds of pieces would there be? Which kind would be the most common?*

Now break up a few of the pebbles. Wrap one at a time in a thick plastic bag, set it on a large rock or curbstone, and hit it squarely with a hammer. Then collect the broken pieces in a paper plate. Let the pupils examine them with a magnifier, and shake them up in a jar of water.

Also let each child examine a pinch of washed sand with a magnifier. *Are the sand grains of different kinds and colors? Are they at all like the tiny pieces of broken pebbles? Might pebbles and other rocks get broken into sand naturally? If so, where and how could this happen?*

MUDPILE MOUNTAIN

Interestingly, some of the same principles involved in the shaping of mountains also apply to the erosion of mudpiles. And so, from a mudpile, a class can get a better understanding of mountains!

Start by having each pupil rule lines, spaced 1 centimeter apart, across both sides of a few Popsicle sticks or tongue depressors. Then have each one crayon bands of color neatly between the lines—red at one end, followed by green, orange, blue, yellow, and purple.

Next, take the class out to a patch of bare ground. Let pupils dig up some of the ground, without large stones, and pile it up to make a "mountain" at least ½ meter high.

Then have the pupils push their sticks, red ends out, into this "mountain" and the ground around it. The sticks should be spaced evenly, each one at right angles to the surface. Only the red, green, and orange bands should show.

Now get volunteers to spray the "mountain" with a watering can, or with a garden hose propped up so that the "rain" falls straight down. Meanwhile, the class can predict what changes will occur. *How will the markers show where erosion and deposition take place? Where will erosion be most rapid? Which will be carried out farther by the run-off water—sand or silt? Silt or clay?* (See "Sorted Stones," page 87.)

Let the activity continue for several days. Have pupils observe any changes that take place and keep a record—perhaps by taking pictures. Point out miniature streams (S), canyons (C), fans (F), lakes (L), and deltas (D), as well as waterfalls and landslides.

Be sure, however, that pupils do not carry this comparison too far. Mountains are not just big piles of mud, but consist mostly of bedrock—and this is not simply washed away. Above all, there is the matter of scale: Compared to this "mountain," a sand grain is a boulder!

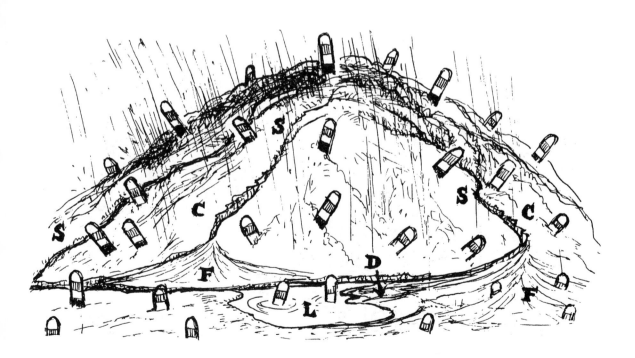

BRICK PEBBLES

Too often pupils merely read and talk about how water "wears away" rocks, without understanding what really takes place. Yet they can easily learn about the *abrasion* of rocks in water, at first hand, by making some brick "pebbles."

Break up a common red brick. Put six or eight sharp fragments, each about 1 inch across, in a strong, thick-walled jar that has a cover. Add water until the jar is half full. Then screw on the cover tightly, and wrap the jar in a strong paper or plastic bag, for protection.

Now ask 10 volunteers to give the jar 100 shakes each—but not so hard as to break it. After 1,000 shakes, have the pupils examine the fragments, observe the water, and scrape the inside of the jar with their fingernails. Pour the water into another jar and let the mud settle. (See "Sorted Stones," page 87.)

Let other pupils repeat this, replacing the water and giving the fragments 1,000 shakes each day.

Have them record the number of shakes and note changes in the fragments and the jar. Save the muddy water.

Is it the water *that wears away the fragments? Or do they* abrade *each other, and also the jar? Where does this sort of thing take place in nature? What becomes of the mud formed from real rocks that are abraded in streams and along beaches?*

A record of the pebble rounding can be made with an overhead projector. Before pupils shake the fragments, have them crayon each one heavily on one side. Then place them on the projector, this side down, and focus their images on a large sheet of paper taped to the wall. Pupils can trace the silhouettes carefully, and cut them out. A new set of cutouts may be made after each 1,000 shakes— always with the fragments in the same position and the projector at the same distance from the wall.

Could pieces of real rocks, such as sandstone or limestone, be abraded in this way? Would adding sand to the jar speed up the abrasion? Let pupils find out.

LOST SOIL

As water from rain and melting snow runs downhill to streams, lakes, and the sea, it carries along bits of the land. This results in the loss of tremendous amounts of one of our most valuable resources—soil. Although soil loss cannot be stopped entirely, we should do everything possible to reduce it.

Pupils can learn at first hand about soil erosion and ways to reduce it by making a few *controlled* tests. In each of these there are two setups, just alike except for one factor. This is the factor being tested.

For example, teams of pupils may make the first controlled test described below. Then they may carry out some of the modifications of it. In each case, two miniature "fields"—different in only one respect—are "rained on," and the amounts of eroded soil are compared.

1 Fill two pans, such as aluminum cake pans, with soil. Pack it down and level it off so that its surface is even with the top edge. Set the pans outdoors on a clean concrete walk. Leave one flat, and tilt the other by propping up one end. Then spray both pans *equally* with a watering can or garden hose, or wait for a hard rain. *From which pan is more soil washed out onto the concrete? Why is this?*

2 Tilt *both* pans of soil, one steeply and the other gently.

3 Fill one pan with sod "scalped" from a mowed lawn, or plant rye grass in it. Leave the soil in the other pan bare, and tilt both pans equally.

4 Tilt both pans the same, and "plow" the soil in them. Have the miniature furrows run up and down on one slope, and crosswise on the other, as in *contour plowing*.

5 Cover the soil in one pan with a *mulch* of dead leaves or straw, and leave the soil in the other pan bare. Tilt both pans to the same angle.

STONE SIZES

Pieces of rock show a great range in size. A class can become aware of this, and can learn the names of the various sizes, by collecting stones and then sorting them.

First ask each child to pick up three stones of different sizes. Then let everyone help to line up all the stones in a row, from largest to smallest. Also help them to make a label for each size, as follows:

Boulders (more than 256 millimeters across)
Cobbles (between 64 and 256 millimeters across)
Pebbles (between 4 and 64 millimeters across)

Should very tiny *pebbles be put in the row of stones? How about sand grains—are they also pieces of rock?* The class may then decide to add these to the collection:

Granules (between 2 and 4 millimeters across)
Sand grains (between 1/16 and 2 millimeters across)

Are there bits of rock even smaller than sand grains? Let the pupils decide, after examining dried mud from a puddle—rubbing it between their fingers and on dark paper. Usually mud contains *silt* and *clay*.

Silt consists of rock particles about as small as the particles in scouring powder. It crumbles readily to a powder when dry. And if some is rubbed between a coin and the bottom of a can held to one's ear, it causes a grinding sound. For comparison, this should also be tried with nothing between the coin and can, and then with sand between them.

Clay consists of rock particles so small that, singly, they cannot be seen without a microscope—and usually not even then! Their minute size makes clay feel very smooth. When moist, the particles stick together, allowing clay to be molded into various shapes. And they stay stuck together when dry.

A labeled collection of stones of various sizes is good to have on hand. Pupils can refer to it, for instance, when examining soil dug up to plant a tree, or when reading about the Mississippi Delta.

SORTED STONES

As a follow-up to "Stone Sizes," pupils can easily investigate how stones get sorted according to size by water. This is a very common and important process in nature.

As part of such an investigation, pupils can do these things:

1 Fill a tall bottle to the brim with water. Then drop in, at the same instant, a sand grain and a small pebble. *Which settles faster?* Check this a few times. Also try sand grains and granules, sticking them to a wet fingertip and touching them to the water. Similarly, compare the settling of sand and silt. Finally, add a little clay. *How long do the clay particles take to settle?*

2 Dig up some *subsoil*, preferably without roots, from beneath *topsoil*—perhaps at a road cut. Drop a handful into a tall jar nearly full of water. *As it settles, does sorting take place? What hap-* *pens when more handfuls are added from time to time? Might the layers often seen in a sand bank have settled in water that was there at one time? If so, which layer is the oldest?*

3 Get samples of subsoil from various places, and compare them by letting water sort them. For fair comparisons, put equal amounts in similar jars and add like volumes of water; then cover the jars tightly, shake them hard, and let them stand. *What differences show up? In which samples, if any, are the rock particles mostly about the same size—for example, mostly sand or mostly clay?*

4 In like manner, compare samples of topsoil from various gardens and fields. *What sizes of rock particles do they contain? How much organic matter is in them?* Usually this settles quite slowly, or else it floats.

5 Measure how far various sizes of stones sink in water in 1 second. (See "Swinging Second-Timer," page 19.) Show the distances on a graph. *Does the shape of the stones make a difference?*

DISTANCES STONES SETTLE IN WATER IN ONE SECOND

SAND SPACE

From past experience, perhaps as suggested in "Pebble Jar" (see page 82), pupils know that stones do not fit together exactly. There are gaps between them. *Are these gaps separate spaces, or do they all connect together? About how much space, altogether, is there in the gaps—between the pebbles in a jarful of pebbles, for example?*

Fill a jar with clean, dry pebbles. *Is the jar* full? *How much space, in all, is there between the pebbles? Is there one-half of a jarful of space? One-third? One-tenth?* Let each pupil make a "guess-timate" and write it down.

Then let pupils check to see how accurate their guesstimates are. They may use a small paper cup or a measuring cup to find out how much water can fit between the pebbles. Of course, they should shake out any trapped air bubbles. As a check, they may measure the water again as they pour it off. Finally, they should measure the volume of the "empty" jar. (*Is an "empty" jar really empty?* (See "Empty or Full?," page 31.)

The amount of space between the pebbles may be expressed as a fraction, such as $3/10$ (as when there are 3 cups of space between the pebbles in a jar that holds 10 cups when "empty"). Or it may be stated as a percentage, such as 30 percent (as when there are 600 milliliters of space between the pebbles, divided by 2,000 milliliters—the volume of the "empty" jar—and multiplied by 100). Incidentally, a quart contains 32 fluid ounces, but many "quart" jars hold more than this. (See "Liters, Quarts, and Quasi-Quarts," page 16.) Therefore, the actual volume of the jar must be measured.

A class will be fascinated by another way of measuring the space between the pebbles. Hold them tightly in the jar, turn it over, and lower it straight down into a pail of water. Then let the pebbles fall out gradually, so that all the air between them stays in the jar. Mark the volume of this air, perhaps by a rubber band around the jar at the exact water level. Place the mark while holding the jar so that the water level is the same inside as out. Then lift the jar out and fill it with water

up to the mark. This much water has the same volume as the air that stayed in the jar when the pebbles fell out. It shows how much space there was between the dry pebbles!

Would the pebbles have the same amount of space between them if they were smaller? If they were larger? If they were more perfectly round? Which would have more space between them— pebbles or marbles? Pebbles or sharp pieces of rock, such as crushed stone from a driveway?

After finding the answers to some of these questions by actual tests, pupils should investigate the space in sand. Sand space is far more important than most of them realize!

Is there as much space between the sand grains in a jarful of dry sand as there is between dry pebbles in the same jar? Let pupils find out. However, since air gets trapped in dry sand when water is poured on it, the second method suggested

above is better than the first. The jar lid may be used to hold the sand in the jar, and to let it fall out slowly when the jar is under the water. Otherwise, pupils may measure the space in sand by adding water from *below*—perhaps through a drinking straw that goes down to the bottom of the sand in the jar.

The space in sand, and also that in gravel, is very important. Water from rain and snow soaks into it and, deep in the ground, this space is usually all filled with water. Countless wells get their water from water-saturated sand or gravel.

In many areas, sand has hardened into *sandstone,* and gravel has formed *conglomerate.* However, even though these rocks are firmly cemented by minerals, there usually is much space left in them. Underground, as a rule, they hold huge amounts of water—and, in some places, natural gas and petroleum.

Pupils can better appreciate that there is space in "solid" sandstone and conglomerate by watching drops of water soak into these rocks. They can also weigh samples before and after soaking them in water, and note the change in weight. For water to soak into dry rocks, what must come out?

The concept is made even clearer by a model that represents sandstone, greatly magnified. To make it, set a berry basket in a pan and fill it with pebbles. Next, thin some glue with water and drip it on the pebbles until they are all wet with it. Keep spooning it over them, and then let it dry. Later, break off the basket. The pebbles will be held together by glue, much as the sand grains in sandstone are cemented by minerals. *Are the spaces between the pebbles all connected? Can pipe cleaners or wire be run through the model? In similar fashion, could water move through* real *sandstone?*

YOUNG FOSSILS

Children find fossils fascinating. They like to hunt for them, and they often bring their finds to school. Whether or not they find any, they should understand what fossils are and how some kinds have been formed. To this end, let them make some "fossils."

First ask volunteers to collect mud, perhaps from a dried-up puddle, and put some in a large pan. Have them add water and take turns stirring with a stick. *Where in a sea or lake may there be soft mud like this? What might stir it up?*

Have pupils keep stirring the mud and adding more until the pan is nearly full. Meanwhile, let each one, in turn, drop in a shell, fish bone, crab claw, starfish, or other seashore souvenir. Or let them drop in shells of fresh-water clams and snails, crayfish claws, and bones—even leaves. Everyone should add something. *How is this like what happens in nature? Where?*

After this, set the pan aside and let the mud dry out. This takes less time in a warm, dry place. Eventually the mud should get hard—a little like what happens when ancient sea-bottom or lake-bottom sediments harden, in time, into rock.

After the mud is completely dry, let everyone take part in breaking it apart to discover the "fossils" in it. Also have them look for the impressions left by shells, leaves, and other things. *Should these impressions be considered to be "fossils"? If so, how many "fossils" can a single shell form?*

As a follow-up, take the class on a fossil-hunting expedition—if fossils can be found in your area. Otherwise, visit a museum to see some.

It is also good to go to a nearby lake or seashore—or even to a mud puddle—to see how fossils may be formed. *How much chance do dead fish, insects, and leaves—or the footprints of birds—have of becoming fossils? What would have to happen to them? Instead, what almost always* does *happen to them?*

MINERALS IN MEMORIALS

An excellent way for pupils to observe the minerals in rocks is to examine memorials in a cemetery, or polished stones in buildings. Be sure to get permission beforehand, and take along magnifiers—*but no hammers!*

Start with a colorful, coarse-grained stone such as granite. *What different colors can be seen in it? Are some of the grains transparent, like glass? Are others* translucent, *like wax, or* opaque, *like pencil lead? Do any shine like metal, or change colors like peacock feathers?*

As a rule, grains that look distinctly different are of different minerals. *Which kind is the most abundant? Are any rather rare? Could a flea walk across the stone on grains of only one kind? Or would it have to hop to get from one grain to another of the same kind?*

Let pupils drip water on the stone and see whether it soaks in. *Do the grains have gaps between them, or do they fit together tightly? Do they seem to be separate bits of rock, like sand grains, which have become cemented together? Or are they more like crystals that have grown from solution?* (See "Sand Space," page 88, and "Reappearing Act," page 112.)

Now let pupils compare the minerals in different stones. They can note whether the various kinds are distributed evenly or bunched up in places. They can look for swirls that suggest movement, as though the rock had once been soft.

Geologists infer that granite and similar kinds of rock, often used for memorials and building stones, were formed deep underground—beneath miles of other rock, under tremendous pressure. (*How many memorials would there be in a stack a mile high, and how great their pressure?*)

At this depth, the rocks were hot and probably soft, if not actually molten. Then, as they cooled, crystals of various minerals grew. And, since then, the great thickness of rock that once covered them has been eroded away, leaving them exposed where they can be quarried.

SEA-SEDIMENT SILLS

In many communities, school buildings have window sills of light-colored stone. These may look like concrete—and some sills are concrete. Many, however, consist of natural rock, a kind of limestone. A great deal of this limestone is being quarried in Indiana.

This Indiana limestone is an ancient sea-bottom deposit. In it one can sometimes see pieces of the skeletons of sea animals. *Can you find any?* The animals used to live in a shallow sea that once covered the area where Indiana is today. When they died, their skeletons sank into the sediment on the bottom, and in time the sediment hardened into limestone. Large blocks of this rock are quarried in Indiana and cut into building stones of various shapes.

What happens when a tiny loose bit of this rock is dropped into a little vinegar? Do bits of other kinds of rock act like this, too? How about a bit of a seashell or coral skeleton? This is a test for "limy" material—more correctly termed *calcium carbonate*—whether in limestone or marble, in seashells or coral skeletons.

ICE-AGE FUN

Children may read that large parts of the earth—perhaps even their home areas—were covered by great glaciers during the Ice Age. To make their reading more meaningful, however, they also need firsthand experiences. Therefore, try these activities with them:

1 Pack snow into a can. *Can you get it to form ice? How does the amount of air in it change?* (See "Empty or Full?," page 31, and "Sand Space," page 88.)

Glaciers are formed from snow that changes to ice, with much air squeezed out.

2 Slide an ice cube over a pane of glass lying flat. *Does it make scratches? What if you put bits of rock under it?*

In many areas bedrock shows scratches made by stones dragged along under former glaciers.

3 Mix sand with snow and pack them in a large milk carton. Cut off the carton and set the sandy ice in a pan indoors. *As it melts, what happens?*

In many places, as former glaciers melted, they deposited hills—often being dug up today for sand and gravel.

ROCK SOLUTION

Limestone and marble are common rocks, much used for building and ornamental stone, for crushed stone, and for making cement and lime. Children should learn to recognize these rocks, and can readily do so.

Both are soft enough to be scratched by a knife. And both give off bubbles with some acids, such as vinegar. (See "Sea-Sediment Sills.") Any rock that does these things is likely to be limestone if dull, marble if shiny or sparkling.

Underground, in some areas, these rocks have caverns in them—the result of slow solution by water containing weak acids, as it seeps through cracks in the rocks. And, once caverns have been formed, water keeps dripping into them in places. Here the rock material comes out of solution again and forms *stalactites* and *stalagmites*.

It is easy to show that limestone and marble can be dissolved. To do this, make a long trough of corrugated cardboard, line it with plastic wrap, and prop it up so that it tilts slightly. Then get a few handfuls of limestone or marble chips. Wash them well with clean rain water and spread them all along the trough.

Now let pupils add clean rain water, drop by drop, at the top of the trough. It will seep through the chips of rock and drip out at the bottom. After several minutes, have someone catch a few drops on a clean pane of window glass.

After the water evaporates, hold the glass up to the light. *Has anything been left on it? If so, where did this material come from? Could it have been in the water to begin with? How can one be sure?*

Does adding vinegar to the rain water speed up the solution of limestone or marble? In nature, water often contains acid—not vinegar, but mostly *carbonic acid*, formed when water dissolves carbon dioxide. (See "Flame Killer," page 72.) In addition, people pollute the air with acids, and these are washed out by rain.

MUD-PUDDLE STORIES

Mud puddles make better books than one might think. Fortunate is the school that has some nearby, and wise the teacher who lets pupils learn to read them!

A partly dry mud puddle is good for a start. *Where did the mud come from? Did the water run in from all sides, or as one definite stream? Where did it flow out—if it did? How high did it get? Are there any signs of lower water levels?*

At another time, perhaps, when the mud is nearly dry, cut out small blocks with a knife. Trim their edges, and let pupils examine them with magnifiers. *Is the mud in layers? If so, how many layers can you count? Which one was deposited first, and which last? Do the layers show how many times it rained?* (See "Sorted Stones," page 87.)

Encourage pupils to keep observing mud puddles. *How are they like lakes? When they dry up, why does the mud crack? What would happen if wind blew sand over the cracked mud? Then what if it rained again, and more mud were washed in?* Pupils can test their ideas by spreading dry sand over cracked mud in a dried-up puddle and looking again after a rain has washed in more mud.

Can you see any signs of animals having walked or wriggled across the mud while it was soft? Have any bones or leaves gotten buried in the mud? Pieces of newspaper? Are there any small round dents made by raindrops hitting the soft mud?

Later, present this problem to the class: *Can you imagine a lake drying up completely? What would be left? From the dried mud how might you tell:*

1 *Which layer is the oldest: the youngest?*

2 *How high and how deep the water got to be?*

3 *Whether the lake had ever dried up before and later became filled up again?*

4 *What kinds of animals and plants used to live in the lake and nearby?*

5 *How many years the lake had been there before it dried up?*

Environment and Conservation

The term *environment* is familiar to children. However, it is applied to so many things so casually, from household climate to the biosphere, that children may be confused about its meaning and unimpressed with its importance. If children are to become thoughtful custodians of their environment, they must learn what constitutes their environment and how they interact with it. Children need to learn about environmental limits, interdependence, and rights. They must develop a sympathy for living things that recognizes that each may have a place not just for exploitation by people, but in a wider sense of environmental harmony. Children must learn to make considered judgments about their actions on the environment, knowing that change may range from slow and controlled to swift and irreversible.

Of the plants, animals, soil, water, and air around them, children should learn that:

Certain organisms and relationships are fragile and will not survive careless treatment.

Most relationships are far more complex than they seem to be at first glance.

Many basic, essential resources are used over and over again, so that each user must be considerate of the next.

Studying the environment may begin with simple observations of nearby, common organisms and objects such as flies, caterpillars, seeds, stones, spiders, twigs, and flowers. Let the children look with magnifiers at structures and movements. Simple equipment such as plastic boxes with built-in magnifiers, or gauze-capped baby-food cages, will aid in this initial examination.

From simple observations will come questions such as, "What's it doing?" or "How can it walk on a web that's sticky?" You may not know, and should not give, answers to such questions. Rather, the children should be encouraged to carry out simple investigations to find the answers. What children find out for themselves often is far more convincing than adult say-so.

Children should be guided in their handling and care of organisms so that injury and waste are minimized and a sympathy for all living things is established. Developing responsibility and consideration for others should be long-range objectives. When children understand that they can destroy in minutes what Nature takes years to develop, then their treatment of the environment may be more responsible and considerate than has been their predecessors'!

SOME IMPORTANT OBJECTIVES

Attitudes and Appreciations to Be Encouraged

The interaction of living and nonliving things is exceedingly complex; an action of one may affect another seemingly unrelated, or a long distance away, or even after a long time.

The natural resources that are available in any practical way to people are limited in amount and confined to a very thin layer of the earth.

No one person or group has any more right, by destiny or decree, to the basic necessities of life than any other person or group.

Some basic resources such as water are used by all living things, and recycled, so that what is used today has already been used many times by forms no longer living.

Because space on earth is limited, and some discarded materials are slow to decompose or are even harmful, people must exercise increasing care in what they discard and accumulate.

"Away" usually is just some other place on earth.

"Take nothing but pictures; leave nothing but footprints!"

How someone else left it for me is *his* or *her* responsibility; how I leave it for someone else is *my own* responsibility.

Some environmental changes can be reversed, but others never can be.

Only people see other organisms as good or bad; for the environment as a whole, organisms are neither good nor bad: they just *are!*

When the environment is affected by something, people are affected, too, because they are a part of the environment.

Skills and Habits to Be Developed

Minimizing waste at home or school by efficient, thoughtful use of materials

Taking care of both personal and public equipment to make it last

Sorting discarded materials efficiently so that items suitable for recycling can be recycled

Declining a carrying container at the store for an item that is already packaged

Deciding not to use aluminum foil when a biodegradable substitute will do

Pointing out alternative points of view, with reasons, when organisms are described as bad or good

Describing the limiting factor that prevents a particular situation from being improved

Exhibiting a sympathy for living things through care of organisms borrowed for study

Exhibiting a willingness to pick up others' litter and to dispose of it properly

Observing carefully and keeping accurate records of changes in one or more objects or conditions in the environment

Facts and Principles to Be Taught

The environment is a complex association that includes organisms, soil, water, air, and sunlight.

People are an integral part of the environment, and they interact with other things in it.

The environment is never truly balanced but is in a state of constant change.

Some environmental changes, such as the growth of lichens, are so slow that they are unnoticed; others, such as earthquakes and forest fires, are very rapid and visible.

People using machines and chemicals can slow environmental change or speed it up.

Some materials, such as aluminum, glass, and certain plastics, take so long to decompose that they may be considered to occupy space permanently.

The half-life of radioactive waste is the time it takes for half of what is left to decompose.

There is no known way to speed the decomposition of radioactive waste.

Recycling ordinary waste can be a useful and nearly self-supporting source of energy and raw materials.

A LITTER BIT LESS!

Ours is a throwaway society. Witness the names of products such as "Chux" and "Flicker," as well as phrases and slogans such as "Chuck it!" "Flick your Bic!" and "One use only." Bottles, cans, plastic, and foil often litter roadsides, picnic areas, trails, and some of the areas around school. If children worked to clean up such an area, they might appreciate how nice their environment could be when unlittered. They, themselves, might also be less likely to litter after having worked to clean up someone else's litter.

Take the class to a nearby park, woods, field, or safe waterway. Divide the class into teams of about four, and give each team a large trash bag. Then assign each team to a specific area to "police." Have them collect and drop into the bag all objects from that area that were discarded by people. One child in each group should keep a record of the kinds of things collected. At the end of half an hour, ask all groups to assemble at one place with their bags.

What are the most common objects and materials in the litter that was collected? How should these kinds of things have been disposed of? Where is the nearest trash container in which they could have been placed? What can be said to people who litter that might make them change their behavior? A week or so after the cleanup, return with the class to the same area to see whether the cleanup deterred people from littering again. *What is your reaction to what you observe?*

Let the class think about and discuss the litter problem in and around the school. *Where is the problem worst? What kinds of things make up most of the litter?* Invite the custodian to share with the class the problem from a custodian's point of view. From the dialogue, try to elicit from the class suggestions for improving the litter problem in and near the school.

Perhaps the class can arrange to set a large container where it will invite trash deposit and discourage people from throwing things on the floor or the ground. Schedule a container check for each day, and rotate the responsibility for emptying it. *Is the location of the container an effective one? How can the container be made more attractive?*

Have the class create some signs or slogans to call attention to the trash depository. See what they can suggest that is forceful yet tactful, and thought provoking without being smart or demanding. *What changes can you notice in the behavior of your own class as a result of this activity?*

THAT DEPENDS!

Write on the board, ''That depends!'' and find out what the class thinks it means. Regardless of what the pupils think, leave the expression there until after the class has given some attention to the following activity. Then review with them the implication of ''That depends!''

Collect some illustrations, dead or living specimens, or models of various organisms that children might characterize as good or bad—perhaps because of a name, appearance, or even hearsay. Some organisms to consider are:

poison ivy (poison oak)	garter snake
dragonfly	tarantula
bat	dandelion
spider	mosquito
meadow mouse	hawk

Starting with one organism, such as poison ivy (or poison oak), ask the pupils whether they think the organism is good or bad, and why. Most children will say ''Bad!'' for poison ivy. And it is true that poison ivy can cause severe dermatitis.

Now ask, *But what color is poison ivy? What causes the green color? What does chlorophyll do in plants?* (It helps them to make food and oxygen from carbon dioxide and water.) *Then, is poison ivy good or bad?* In that it carries on photosynthesis, it is good. It helps to put oxygen into the air. While it may be poisonous to touch, it is helpful in improving the quality of the air we live in.

Poison ivy often grows in waste places, sometimes on soil that might be eroded if the roots of plants did not hold it in place. The roots of poison ivy are fibrous and help to hold soil in place. *In view of its roots, is poison ivy good or bad?* That depends upon where it grows, whether people are near it, and what might happen to the soil if poison ivy were not there. Discuss with the class poison ivy's potential for cover for small animals, how it carries on photosynthesis, helps hold the soil in place, and still can poison people. *Is it good or bad?* That depends!

Or let the class consider a dragonfly. Let them read about dragonflies, observe them flying, and find out what they eat (mosquitoes and other small insects). Children might think that dragonflies are good because they eat mosquitoes—except that dragonflies have long ''stingers.'' But that long stinger cannot sting! In females, it is for laying eggs. In males, it is for holding the female during mating.

Let the class collect one or more dragonfly nymphs at a local pond and keep them in a shallow pan of pond water. Feed them mosquito wigglers, bits of earthworm, or even the bodies of houseflies held in tweezers under water. Once the children see the lower lip of a dragonfly nymph flash out to seize prey, they will be fascinated. Eating mosquito wigglers! Dragonfly nymphs must be good! *For whom? Mosquitoes? Other organisms such as small fish that depend upon mosquito wigglers for food? Bigger fish such as bass that, in turn, eat smaller fish that depend upon mosquito wigglers? And so, are dragonflies good or bad? What makes* ''That depends!'' *an acceptable answer?*

Put a three-column chart on the board. In the first column, list some organisms such as those named above. Above the second column write ''Good (helpful).'' Above the third write ''Bad (harmful).'' Let the class find out all it can about how each organism can be good and bad. Let the children consider *for whom* or *for what* each is good or bad, and how. When one of the class later describes a particular organism as bad, ask ''For whom?'' Encourage a ''That depends!'' kind of response from the children, and the formulation of reasons for such an answer.

If your class can, and wishes to, write or act out roles, let them describe an organism's life from *its* point of view. What might it be like to be:

A spider who must simply wait until food comes by?

A hawk whose young need food to live and grow?

A mouse living in a meadow next to a woods where a hawk nests?

Children's sympathies may not always be with the prey. That depends!

DISAPPEARING ACT

Of all that we discard, what things decompose to become part of the soil, or food for organisms? What things are least likely to decompose or disappear? Some simple tests and observations may help to answer these questions, and thus help develop in children a care that is badly needed in today's throwaway society. For example, children can experiment with, and see what happens to:

Small squares of aluminum foil, paper, cloth, can metal, or plastic
Iron nail, screw, nut, or washer
Orange peel, banana skin, slice of bread, or peanut shell
Ball-point pen, pop-top ring, bottle cap, rubber band, or Band-Aid

Let each child choose one such thing, and then decide what conditions to impose on it, such as:

Bury it in moist soil.
Try, with an adult, to burn it outdoors.
Put it in water.
Freeze it.
Leave it on a table in the sunlight.
Seal it in a small plastic bag.
Put it in vinegar, boric acid, or baking soda.

After a week of repeated treatment and observation, have the class report what happened. Some things may have started to decompose or decay. Some may have burned to an ash, and some may only have charred. Some may have begun to rust or change color. A few may not show any visible change, in spite of what was done to them. A piece of aluminum, for example, may be almost free of change, even after a week of experimenting.

Some children may wish to weigh their objects to see whether there was a gain or loss of mass, and how much. They may also want to check the color, smell, and feel of things.

When all the evidence is in, let the children decide for themselves:

1 *What discarded things are most apt to change, even in a short time?*

2 *What things should we be most reluctant to discard because they do not change much at all?*

3 *What things are most likely to be untouched by heat? By moisture? By other common chemicals?*

4 *Of all the things we throw away each day, what ones probably cannot be used by plants? By animals such as flies, beetles, and other small creatures? What things probably should be taken to a special place for recycling because they do not disappear?*

WHERE IS "AWAY"?

One pressing environmental problem is the accumulation of waste. In time, we may be hard-pressed to find new dumping places for both solid and liquid waste, to say nothing of the waste sent daily into the atmosphere. If children are to be made sensitive to their individual contributions to this problem, so that they will change their habits and alleviate it, they must first be made aware that "away" is only someone else's backyard.

Show the class a piece of unused chalk. Let pupils discuss how they know it is unused. Perhaps they can sketch it, either individually or as a class. They might project its shape on the board, and then trace around it, marking the outline "Save!" so that they can compare it with a later outline. Then let each child write on the board with this piece of chalk so that the end is well worn.

Ask: *How can you tell that this chalk has been used? Where is the chalk that once was part of this piece?* At this point have the children use a magnifier to look at the marks on the board. They should see that the marks are not smooth but are pathways composed of bits of chalk. *What can we do to those lines on the board?* The children's response almost always is, "Erase them!"

When the children have erased the writing on the board, call their attention to where the chalk is. Let them look through a magnifier at the bits of chalk on the eraser. When all have seen them, have someone clap two erasers together in the beam of a projector shined across the room. Bits of chalk will show up well as light from the projector bounces off them. (See "Orange-Peel Mist," page 61.) This is the same principle as dust showing up in a shaft of sunlight shining into a darkened room.

It is best to let this chalk dust fall onto sheets of black paper placed on the floor of the classroom. Then the children can examine them once more with the aid of a magnifier. While they look at the bits, review with the class how the bits got there: piece of chalk—board—eraser—floor.

Let the children dump the bits of chalk from the paper into a wastebasket. Then go as a class with the wastebasket to the large trash container that awaits the coming of the trash truck. If possible, let the class watch as the container is emptied into the truck and the bits of chalk are taken away. In some schools it may be possible for the class (in a school bus) to follow the truck to the dump.

Some questions that you might ask are: *Is "away" for the chalk a definite place? Is that where the chalk will be tonight? How about the chalk we erased last week? Last term? What other things might be there, too?*

Follow this with a discussion of, and an examination of, worn heels, worn tires, worn clothing, and worn pencils or crayons. *Where is "away" for the bits of heel that are no longer on your shoe? Where is "away" for the milk carton from today's lunch? Where is "away" for the cap from a bottle of pop? For the pop-top from a can of soda? Where do you think "away" might be for the dirt that is taken from your clothes in a washing machine? For the grass clippings mowed from people's lawns?*

Children's drawings, taped accounts, and class discussions are excellent grist for developing lessons that will provide an efficient learning climate for the concept "away."

A WEB OF LIFE

An important concept relating to the environment is *interdependence*. One dramatic way to illustrate interdependence is to have the class make around-the-neck signs for themselves from oak tag and string, and then proceed to play a game as follows.

From a box containing the names of common organisms, materials, or prominent structures in the environment, let each child pick one. Examples to include are:

robin	spider
mouse	housefly
water	corn
grass	clover
sunfish	sun(light)
cow	dog
soil	earthworm
human being	apple tree
house	honeybee

As the pupils pick the name of individual things, have them write the names on their signs and wear them. There will be plenty of laughter at first, as children are identified with various names. As the game progresses, however, interest in interdependence will grow and silliness will wane.

Get a ball of lightweight string, or readily visible thread, and scissors. Then take the pupils to a place where they can stand in a circle (outdoors on a nice day). Start with one name, such as Soil, and cut strings to connect the child so labeled with every other child whose sign indicates a direct dependence upon soil. For example, Soil might be holding strings running to Grass, Apple Tree, Clover, Corn, and Earthworm. As each connection is made, discuss with the class what that connection, or dependence, really is. Discuss, also, how it may be an *inter*dependence, not just a one-way dependence. For example, discuss how grass needs soil to grow, but cuttings decay to become part of the soil. Soil is made up of bits of decayed grass, as well as many other particles.

Name by name, go around the circle, letting each child hold strings connecting all others dependent upon, or interdependent with, him or her. In time, the class will see a substantial web of life that shows some of the interdependence among organisms and materials in the environment.

Now, to demonstrate what things might be affected by the removal of one, let Soil drop its strings. *What organisms would be seriously affected by the removal of soil? If Soil picked up the strings, only to have Water drop its strings, what would be affected? What organisms would be affected if Sun(light) and Water both dropped strings?* Almost everything!

The web of life shows only direct connections. But this simplified dramatization might be supported and extended by a discussion of indirect interdependence. For example, a cow depends directly upon grass for food. A human being depends upon the cow for milk. So a human being depends indirectly upon grass for food. The human being takes care of the cow, and the cow helps to fertilize the grass. So indirectly, human beings and grass are interdependent. Find out what complex and many-linked chains of interdependence the class can describe. Describing such interdependence, no matter how remote the connections, will help children to visualize the complexity and the relatively fragile character of their own environment.

LIMITING FACTOR

Some environmental changes are either slow or do not take place because something needed for change is limited. Plants may grow slowly because water or light is limited. Birds of one kind or another may be scarce because the number of acceptable nesting sites is limited. Meager funds may be the limiting factor in not getting new books for the classroom.

Children should learn what *limiting factor* means as it applies to the environment, because as adults they may have to deal with limiting factors. They will need to recognize a limiting factor in order to take steps to bring about a change.

There are several places for children to observe limiting factors. One is in a supermarket. Take a class on a field trip to a nearby supermarket. There have the class observe such things as:

The number of shopping carts
The number of check-out counters
The number of check-out clerks on duty
The grouping of goods on shelves
The packaging and cost of selected items

At the time of the visit, what seems to be the limiting factor in how quickly customers make purchases? If people are standing in line, is the lim-iting factor the number of check-out counters, or check-out clerks? Is the grouping of goods efficient? If not, what regroupings might improve shopping?

Let the class decide what *one* condition more than any other seems to slow the customers from completing purchases. That is the limiting factor. *What can you suggest as a way to remove this limiting factor? Then, what would become the new limiting factor? When one limiting factor is removed, is there always another?*

Or, let your class observe another class at the school cafeteria. *What is a limiting factor that makes serving slow? If that limiting factor were removed, what would the next limiting factor be?* With the permission of the school principal, and whatever other administrators are needed, perhaps the class can try an alternative way of serving food, then study the results to see whether their alternative plan really worked.

Another suggestion is to let the class visit a nearby park, pond, woods, or playground. Select some group of organisms such as trees, turtles, woodpeckers, or grass, and let the class decide what must be the limiting factor for the growth and spread of that group. From observations about limiting factors in markets, children should be able to identify some limiting factors affecting growth in a community.

CAN FLATTENER

One problem caused by discarded, but unflattened, containers is simply that they take up much-needed space. Discarded containers are mostly air. If cans and cartons could be squeezed tightly, they would take up only a fraction of the space they normally occupy in trash cans and in dumps.

As practical exercises in science and mathematics, let the class determine the volume of various metal, plastic, and paper containers, both when expanded, and when compressed as much as possible. Some directions for such investigations are:

1 Measure the diameter and height of cylindrical containers. $V = \pi \times (D/2)^2 \times H = 3.14 \times (D/2)^2 \times H$. How would you find the volume of a tapered container such as a cottage-cheese carton? (Hint: Use *average* diameter.)

2 Measure the length, width, and height of rectangular boxes. $V = L \times W \times H$.

3 Set a pan on a level surface. In it stand a large jar or pail. Fill the jar to the rim with water. Then press one of the containers down into the water, catching the overflow in the pan. The water that overflows should equal the outside volume of the submerged container.

4 To find the inside volume of the container, fill it with water and then empty it into a measuring cup or graduate. Or, weigh the "empty" container. Then completely fill it with water, and weigh it again. The net mass of water, in grams, equals its volume in milliliters.

5 To find the inside volume of a cardboard box, line the box with foil or plastic, pressing the lining tightly to the side of the box. Fill the box with water, and then pour the water into a measuring cup or graduate. *What might be sources of error in this measurement?*

Now have the children squeeze, fold, or pound the containers as flat as they can. When the container material is in the smallest space possible, let the class check the volume once more. One way is to submerge completely the flattened container, as in number 3 above, and catch the overflow. The overflow can be measured in a measuring cup or graduate, or it can be weighed. *Why does weight of the displaced water indicate volume? For a particular container, what should cause the difference in measurement between numbers 1 and 4, above?*

From the measurements made by the pupils, let them determine about how much of the total volume of each container is in the material of the container. *About what percentage of the container space would be saved by flattening?*

Finally, hold a class can-flattener contest. See who can come up with the most ingenious and effective can flattener. Some might devise a massive object that would fall and flatten bottomless cans. Others might arrange a lever to flatten bottomless cans. Although these devices may never be put to widespread use, the creativity, fun, and practical science principles in the contest make it well worthwhile.

HALF OF A HALF OF A . . .

As the world need for energy increases, and fossil fuels become scarce and costly, our country and other countries may be forced to turn to nuclear fuels for generating electricity. This means that radioactive waste will be a problem with which children must deal as adults in an energy-shy future.

A characteristic of radioactive waste that causes a problem is that its rate of decay (called its *half-life*) cannot be changed. Whether the waste is contained or free, solid or liquid or gas, in combination with other substances or pure, its half-life remains constant. The half-life of three important radioactive wastes is:

Krypton 85	10½ years
Iodine 131	8 days
Strontium 90	28 years

Children may be only vaguely familiar with some of the names of radioactive substances, such as strontium 90 or iodine 131. They are probably even less familiar with what these potentially dangerous substances do in the body. The pattern of radioactive decay, however, is the same for all radioactive substances and for that reason is important to understand. Besides, it is an interesting mathematical relationship in itself: half of a half of a half . . . forever.

Give each child a sheet of newspaper to cut to the size of his or her desk top. Along one edge, have the children label this starting sheet, "Radioactive! Half-life: 10 years." Then have each one write today's date on the paper. Pretend that this material was created today and that it is now starting its radioactive decay. At the end of 10 years, only half of it will remain. To give the class a time scale that is reasonable, pretend that each calendar day represents 10 years of time.

The next day, have each child cut his or her paper in half and discard one of the halves. Label what is left, "Radioactivity remaining, 19--" and write the date 10 years from the start.

A day later, have each child cut the remaining paper in half, discard one of the halves, and write on the remainder, "Radioactivity remaining, 19--" and write the date 20 years from the start. Do this each day (each 10 years) from the day you begin this activity, and continue it for a week or more. *At the end of a week (seven decades, or 70 years), how much radioactivity is left? How long would it be before none was left?*

The class may wish to graph this relationship on the board, using some other example such as "Walking half the remaining distance each day." Pupils can measure the length of the classroom, and then make a graph showing the distance covered if half the remaining distance is covered each day. *In how many days would three-quarters of the distance be covered? How long would it take for* all *the distance to be covered?* To go half of what is left each day means never to get there! It is this relationship that is important and interesting about radioactivity.

GOODNESS! BOOK OF RECORDS

Children enjoy learning about and communicating world records in sports, human endurance, and dimensions of objects. They know the name of the tallest mountain, the biggest animal, and the batter with the highest percentage or most home runs. They can have a good time, too, finding local records in their own environment—the biggest, smallest, slowest, fastest, strongest, or hardest.

Introduce the class to some twigs of common, local deciduous trees. Show them how to tell the age of a twig, and how much it grows from year to year. Then let them search the neighborhood for a twig that shows the most growth for a single year. Caution them not to bring in the twig, but merely to measure it and describe where it is. (If they get permission from the owner, and your own, they might bring it in for verification.)

Does any twig in the neighborhood grow as much as a meter in one year? More? How much did the slowest-growing twig elongate in one year? How many times as long did the fastest-growing twig grow? What kind of tree was each?

Have the children look for the largest tree leaf they can find. And the largest weed leaf. Let them discuss what is meant by largest. *Most massive? Largest area? Longest?* If they mean area, let them decide how they will find the area of an irregular surface such as a leaf. Some may decide to trace the leaf on a paper of standard thickness, cut out the shape, and weigh it. Others may trace the leaf on squared paper and count squares.

Finding the speediest animal may present a problem. A discussion of how such an animal might travel, and how its speed might be determined, may be productive and may result in some interesting investigations and subsequent measurements. Perhaps some of the children can bring in pets for the day and try to determine their speed over a measured course. As with finding the largest leaf, what the class gains from discussion may well be worth the time and trouble involved.

Some other things to look for:

1 Largest insect. *Largest wingspread? Longest body? Heaviest?*

2 Most numerous animal. *In what area? As shown by sampling? Total count?*

3 Largest earthworm. *Heaviest? Longest? Biggest around?*

4 Most eggs in a bunch. Open a spider egg-mass and count the eggs. Count the cells in one tent-caterpillar egg-mass. *What other egg-masses or nests are there? Where are they?*

5 Most seeds. Count seeds in some pods, flower heads (such as dandelion), fruits (such as melons, grapes), and a cattail (which may have up to a million seeds!).

My GOODNESS!

MAKING DO WITH OLD

A characteristic of modern society, which leads to some environmental problems, is planned obsolescence. New, "improved" models keep replacing old but still workable ones. Styles, colors, and materials change constantly to entice children and adults into buying. Buyers really do not need many of the new things, but advertisements are persuasive and norms are influential. "*All* the kids have them!"

To try to redirect the "need" for new things, and thus diminish the rate at which discarded objects accumulate in the environment, ask the children to bring in some objects that are old, rusty, or out of style—objects that they think should be replaced with new ones. Some suggestions are:

roller skates	tools that are rusty
older-model	or need handles
skateboard	old bicycle or
wooden toys that	tricycle
need fixing	rubber boot with tear

Hold up one of these for the class to examine. Invite some discussion about the object. *Does it work? Can it work? How might the wear or damage have been prevented? Where would the object have gone if discarded? What would have become of it there? How would it have interacted with the environment if it had been discarded?*

Now let the class consider how the object might be restored. *Could it be lubricated? Sanded? Glued? Painted? Straightened? Some part replaced?* When the pupils have made some suggestions for each of several objects, let them try out some of their ideas and observe the results. Perhaps by putting their creativity to work, pride in workmanship may become pride of ownership, and thus temper the urge to discard.

In addition to fixing up objects of seemingly limited value, children can have fun and find challenge in constructing useful things out of apparently worthless discards. For example, small metal juice cans can be hammered flat, little by little, from end to end, to make attractive medallions for a necklace. Other metal cans can be painted to make attractive pencil holders. By pricking small drain holes in washed TV dinner trays, the aluminum trays can become feeding trays for birds. *What other creative ideas can you suggest for everyday discards?*

Some discarded items cannot be fixed or recycled very well, but they can be used for study in the classroom. For example, a wall switch that no longer works might be disassembled by the class to see how it is constructed. A flashlight cell that is worn out might be cut in two so the class can see what is inside it. Old things, worn-out things, broken things, useless things—all may have more value for classroom investigation and creative art than might be expected. Try them and see!

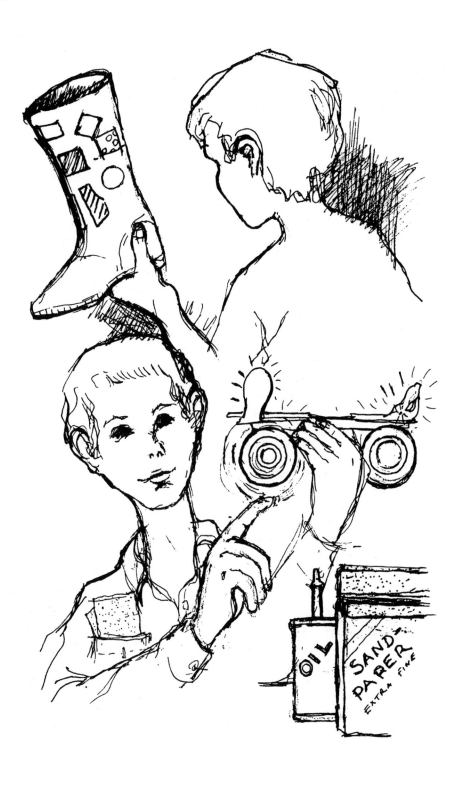

WISE WATER USE

Children often hear about water shortages and may even be affected by them. As the population grows, water shortages will become more common, and obtaining sufficient water for daily needs will be an increasing problem. Since their generation will have to face the problem, pupils should find out how much water is used in daily living and learn to use it wisely. To help them do so, have each engage in a weekend or vacation-period cooperative venture to study his or her family's and community's use of water, as follows.

First, stick a strip of waterproof adhesive tape vertically inside a clean plastic pail. Then *calibrate* the pail by filling it with water, 250 milliliters at a time, and making pencil marks on the tape to indicate liters and fractions of liters.

Next, remove the cover from a toilet tank and mark where the water level comes. Then close the shutoff valve in the water pipe just beneath the tank. Flush the toilet and refill the tank, using the calibrated pail. Record how much water it takes. Replace the cover, turn on the valve, and place a card and a pencil on the cover for members of the family to record each time they flush the toilet. Do this for each toilet if there is more than one in the house. *How much water is needed to flush a toilet? How much is used for this purpose by the whole family during one 24-hour day?*

Put another strip of adhesive tape vertically on the inside of the bathtub, near the drain. When you take a bath, run the water into the calibrated pail first, a liter or two at a time, and then fill the tub, making a mark on the strip of tape for each liter added. *How much water is used, on the average, for a bath?*

To find out how much water is used for a shower, stopper the tub before the water is turned on and let the water collect in the tub until the shower is finished. Then record the volume as indicated on the calibrated strip of adhesive tape. *Which takes more, on an average, a shower or a bath? How does the volume compare with that needed to flush a toilet? How much water does the family use per day for bathing and showering?*

To measure the water used for such things as washing a car or watering a lawn, turn the hose or sprinkler into the calibrated pail and record the length of time to collect, say, 5 liters. *How much water does the hose or the sprinkler discharge per minute? Per hour?*

Also try to measure how much water is used for cooking, for washing dishes, and for similar daily tasks. If it is difficult to measure the water used in a washing machine, check the book of instructions that came with the machine, or call the appliance dealer to see if the information can be supplied (1 gallon = about 3¾ liters).

To find out how much water the family has used in a single day, add together all the amounts recorded for all uses, or read the water meter at the beginning and at the end of a day. It will in-

dicate the water used in terms of cubic feet (1 cubic foot = about 7½ gallons, or about 28⅓ liters). *How much water is used by the family in one day?*

When each pupil has completed a record of how much water his or her family uses, let the children add their records together. *How much water, on the average, does each pupil require for normal daily activities? How much, on the average, does each family represented require?*

Finally, let the class find out from one or more local industrial plants how much water they use each day, on an average. Then, by dividing this amount by the number of people employed, pupils can determine the average daily water consumption per employee. *How does the industrial average per employee compare with the home-use average per person? What kinds of industries use the most water? Which industries reduce, or could reduce, the amount by recycling the water?*

The supply of fresh water available for human needs is limited. At the same time, the use of water for both home and industrial purposes keeps increasing. Hence, drastic reductions in its use or some cheaper ways of obtaining fresh water from sea water will be needed if prolonged shortages are to be avoided. To prepare to face future water problems, the pupils of today should think about ways in which water can be conserved. *How does water seem to be wasted at home? At school? In other ways in the community? What are some reasonable ways to cut down on the amount of water used, without affecting ordinary sanitation requirements?*

REAPPEARING ACT

Most pupils have seen sugar and salt dissolve and apparently disappear in water. Probably few have investigated to see what happens to these substances when the water evaporates. *Do they evaporate, too, or are they left behind? If left behind, are they changed to something else?* Some simple activities will help determine the answers.

Set out some labeled containers of common chemicals such as salt, sugar, baking soda, alum, and epsom salt, with a plastic spoon in each. For each group of five or so pupils, provide paper cups, a medicine dropper, a magnifier, a sheet of black paper, and some clean microscope slides or a small pane of window glass. Then let each group proceed as follows.

Put ½ teaspoon of salt in one of the cups. Dump a few grains onto the black paper and examine them carefully with the magnifier. *What shape do they have? What do they taste like?* (Note: These chemicals are harmless to taste in small amounts.)

For each chemical to be tested, label a slide. Set the slide horizontally in a warm place. Next, add 1 teaspoon of water to the salt in the cup, stir with the medicine dropper until the salt dissolves, and then make a puddle of a few drops on the properly labeled slide. Rinse the dropper and repeat the process with the other chemicals.

In the morning, check to see if any drops have left a *residue. Does any of the residue resemble the original substance?* Test each residue for taste. *Which ones taste like the original?* Watch through the magnifier as you add a drop of vinegar to the residue from baking soda. *Does anything happen like what you saw in "Flame Killer" (page 72)?*

Many chemicals that seem to disappear in water reappear when the water evaporates. Examples are spots on windows and on glasses that are not wiped dry. Residues that accumulate over a long time in caverns or under concrete bridges may leave hard projections such as stalactites and stalagmites. Residues also accumulate in teakettles. *Why do you think this might happen?*

Forces and Motions

Children are continually having experiences with things that move—with balls, hammers, automobiles, creeks, and wind, to name but a few. During their lifetimes they will continue to use, enjoy, and avoid moving things. Consequently, they should have a basic awareness of motions of various sorts, and of changes in motion. And they should understand that forces are responsible for changes in motion.

Since forces and motions can often be controlled and measured, the principles that apply to them are among the best understood in science. Tested countless times, these principles seem to hold true everywhere on earth, and also in space—as shown by accurate predictions of the motions of planets and their moons. For explaining and predicting all sorts of actions, from the batting of baseballs to the orbiting of satellites, these principles are invaluable.

In spite of all that is known, however, some false notions persist, and these handicap people in their comprehension of the world around them. One example is that, even if there were no friction, swings and merry-go-rounds would have to be pushed or pulled to keep going. Another is that passengers in an automobile are thrown backward when it starts, forward when it stops, and sidewise when it turns—whereas, like everything else, their bodies merely tend *not* to change motion. Still another is that the earth pulls on a rock, or a hammer pushes on a nail—when, in each case, *both* objects act on one another, with *equal* force!

With today's emphasis on speed, and the tragic toll to which it often leads, it is imperative that young people acquire a sound understanding of the principles of forces and motions. Furthermore, it is important that they develop the realization that these principles, unlike man-made laws, cannot be ignored without inevitable consequences. Increased attention to these aspects of education is bound to lead to a safer and more pleasant life for everyone.

SOME IMPORTANT OBJECTIVES

Attitudes and Appreciations to Be Encouraged

Our understanding of the world around us depends, in part, on our recognition of such fundamental physical factors as force, motion, and mass.

On the basis of many observations, scientists have extracted general principles that describe how forces act and how they affect motions.

Physical principles have countless practical applications, among them tools and other machines.

By mastering relatively few basic principles, one can explain many things that otherwise seem puzzling, such as curious cases of balance.

The same principles involving forces and motions seem to apply everywhere on earth, as well as on the moon and Mars—in fact, throughout the universe.

Events, in science, are not haphazard; thus, one is able to repeat an experiment many times and obtain the same results each time, so long as the conditions are not changed.

Some basic concepts relating to forces and motions are still not completely understood; examples include the concepts of mass, inertia, and gravity.

Many factors in our environment, such as friction, speed, and inertia, have both advantageous and disadvantageous consequences.

For fair, or valid, tests—of friction, for example—the situations being compared must be alike in every respect but one.

Physical principles or laws, unlike man-made laws, seem to be unvarying, and always to lead to inevitable consequences; thus, a crashing automobile causes the same results whether a police officer is present or not.

People commonly have mistaken notions concerning forces and motions—for example, that mud is *thrown* off a spinning wheel, and that the winning team in a tug-of-war is the one that *pulls* harder.

High speeds are dangerous because they make sudden changes in motion possible, and such changes often cause great forces that lead to damage, injury, and death.

Sharp-edged and sharp-pointed objects are hazardous because they enable large forces to be concentrated on small areas.

In trying to explain something that puzzles us, we should suggest possible explanations, or hypotheses, and then check these out by observation and experiment.

Skills and Habits to Be Developed

Identifying pairs of pushes and pulls, and noting the effects that these forces have on things

Determining the strength of forces in various ways, including weighing—a means of measuring the force we call *gravity*

Pointing out ways in which forces are made to act on larger or smaller areas, and explaining the advantage in each case

Drawing sketches to show the direction of motion of various objects, such as balls, hockey pucks, and the taillights of automobiles

Describing changes in the speed and direction of motion of objects, and relating these changes to forces acting on the objects

Estimating the relative mass of objects from the ease with which the motion of these objects can be changed

Comparing the amount of friction between various objects and materials, and showing the differences by simple graphs

Carrying out fair, or valid, tests—as of rolling versus rubbing friction—and pointing out weaknesses that could affect the validity of the results

Using tools such as screwdrivers, pliers, and hammers properly, so as to get maximum effectiveness of the forces applied to them

Giving illustrations of everyday applications of physical principles, such as those of inertia and of action and reaction

Suggesting hypotheses to explain events—for example, why a leaning carton does not topple, or why water drops fly off a wet rag that is whirled around—and testing these hypotheses

Measuring and expressing distances and masses in metric units, such as centimeters and meters, grams and kilograms

Reading, with understanding, such terms as *axle, force, friction, gravity, hypothesis, inertia, mass, pulley, reaction, turning effect*

Facts and Principles to Be Taught

A push or pull is necessary to make something start to move, and to change the direction or speed, or both, of something that is moving.

A force may be made to act on a larger area, as in the case of a thumbtack, by the head; or on a smaller area, as by the point.

An object falls or topples when its center of weight is not supported and is, therefore, able to move to a lower position.

Whenever two things touch there is friction between them, and this tends to hinder any motion of one thing past the other.

Friction between two objects is greater when they rub against each other than when one of them rolls past the other while they are in contact.

The turning effect of a force depends, in part, on how far the force acts from the center of turning—the axle or pivot—of the object that is being turned.

Ropes and pulleys are commonly used means of changing the direction of pulls, as well as of multiplying their strength.

A moving thing keeps going straight unless it is acted upon by a force that causes it to change direction.

When an object such as a ball on a string is whirled around, it is kept in a circular path by an inward pull on it that acts toward the center of whirling.

Forces always act in pairs; a force acting on one thing, such as a baseball, is accompanied by an equal and opposite force acting on a second thing, such as a bat or the hand of a thrower or catcher.

How easily something can be made to start moving or otherwise change its motion depends on the amount of "stuff" in it—its mass.

Weight—although it is a force, not mass—depends on mass; in fact, the weight of something, such as an automobile, is the most-used indication of its mass.

Objects firmly fastened together or held together by friction—such as two pieces of wood glued or clamped together—act as one object, with combined mass.

PUSH OR PULL?

All around us forces act on things, often making them move. Some of these forces are pushes; others are pulls.

To emphasize the difference, lay a block or stone on a sheet of paper and make it move by pulling the paper. *Can anyone make it move by pushing* the paper? Let pupils try. Then roll the paper into a tube, tape it to keep it from unrolling, and let them use this. Also let them stand the paper tube on edge and rest the block or stone on it.

Now lift up one side of a table. *Is this force a push or a pull? If many persons stood around the table and each one used just* one *finger, could they lift it? Could they lift it by pushing or pulling with paper?*

After pupils find out, have them make lists of some other pushes and pulls. *Which ones make things start moving? Which make moving things stop? Which make moving things change their speed or direction, or both speed and direction?*

Does every *force cause a change in motion? To push or pull on each other, must things touch?* (See "Needle Poles," page 149, and "Kinds of Charges," page 158.)

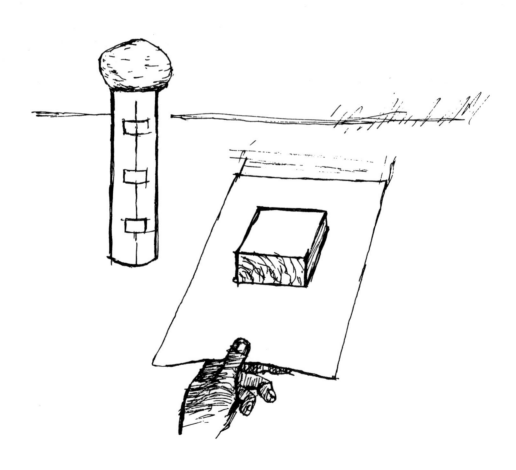

FORCE SPREAD

A force may be weaker than the pull of an ant—or stronger than the push of a bulldozer. What a force can do, however, often depends not only on its strength, but also on how spread-out it is. In short, the size of the area on which the force acts may make a big difference; the force has more effect when concentrated on a small area.

To show this, have each pupil push a pencil into modeling clay or corrugated cardboard—first its point, then the eraser end, and finally the side. *In each case, what happens to the clay or cardboard?*

Why is there a difference, even if the pushes are all of equal strength?

Similarly, pupils may test the edges and flat sides of rulers, and the points and heads of nails. They may also compare the effect of equal pulls on thread and on rope, and on narrow and wide rubber bands.

Then have them make a list of things that cause forces to act on small areas, and one of things that cause forces to be spread out. The first list may include the point of a needle and the edge of an axe; the second, the head of a thumbtack and the feet of a duck.

HOW COME?

It is interesting to observe the reactions of young children when they are confronted by the unexpected—by something that seems contrary to experience. One such thing is a carton that they expect to topple, but that does not do so.

Set up a carton like this before the pupils arrive. Weight it with a lump of modeling clay stuck inside one corner. Then place it on a high shelf or cabinet, with the unweighted part extending far out over the edge. Or prop it up with a block so that it leans far over without falling.

Say nothing about the carton; wait until the children notice it. *Why does it not topple?* Let them suggest explanations—before looking inside it. These explanations constitute *hypotheses* that they invent to account for the unusual situation. (It may be possible to record their comments and ideas on tape, to play back after they have looked inside the carton.)

Another approach is to let pupils, instead of talking about the carton, try to rig other cartons to act in the same way. They should do this without first looking inside the one on display and, ideally, without help. Have many lightweight cartons on hand, and a variety of such things as clay, blocks, rubber bands, nails, tape, string, pebbles, and magnets.

The carton does not topple because it and the lump of clay inside are essentially a single object, with most of its weight far off center. In fact, one may imagine *all* its weight to be concentrated at a single point—near, or even inside, the clay. This point may be called the *center of weight.* (The usual term is *center of gravity,* but to a child the concept of *weight* is probably simpler and more direct than that of *gravity.*)

So long as its center of weight is supported and cannot fall, the carton-and-clay object stays put. *What happens, however, when its center of weight is not supported and can move to a lower position?*

FRICTION CUBES

Whenever materials touch each other, there is friction between them. How much there is depends, in part, on the materials. Teams of pupils can discover this by making fair tests with friction cubes.

First, at a lumber yard, have a four-by-four sawed accurately into cubes—one for each team. Also get a smooth board for each team, and some glue or liquid cement. Then ask pupils to bring in pieces of various materials such as leather, rubber, plastic, felt, aluminum, sandpaper, and linoleum.

Next, have each team cut squares of different materials to fit the faces of a cube, and glue or cement them on smoothly. Their edges should be trimmed so that the cube rests on only one material at a time. One face of the cube may be left uncovered, to provide a surface of bare wood.

Now team members can take turns making tests. Each should set the cube on a board, and slowly raise one end of the board until the cube just starts to slide. Then the height of the raised end indicates how much friction there is between the board and the material in contact with it. If the board rests on a chalk tray, this height can be marked on the chalkboard and then used in drawing a bar graph.

In this way, each team may make fair tests of the friction between the board and six different materials. *Why are the tests fair as long as the same cube and board are used?*

A team may also:

1 Test the friction between the cube and other surfaces. These may include a painted board, a sand-covered board, a rubber mat, and a sheet of metal.

2 Compare the slipping of waxed surfaces of wood with that of unwaxed ones. For this a block and a board—each waxed on one side only—work well if the block is weighted by a brick.

3 Investigate the gripping of shoes, sneakers, and rubbers on various kinds of floor surfaces. *How can the tests be kept as fair as possible?*

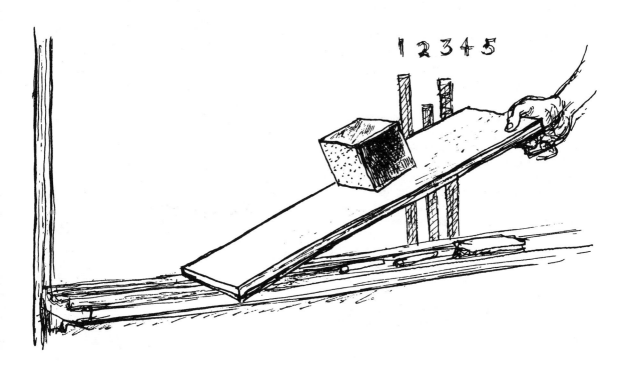

RUBBING VERSUS ROLLING

Whenever one object rubs against another, friction between them tends to retard the motion. Such friction is often a hindrance, resulting in a waste of energy. One way to reduce it is to substitute rolling for rubbing—as pupils can easily discover for themselves.

Put some heavy objects in a box, such as a shoebox, and fasten a slender rubber band to one end. Let pupils slide the box along a table by pulling on the free end of the rubber band. Then let them set the box on several drinking straws, to act as rollers, and again pull on the rubber band. *What causes the difference in how hard one must pull?*

For a valid test, the situations that are being compared must be alike *in every respect but one. Is this true if the box* rubs *on varnished wood, but* rolls *on plastic straws? How can the test be made fairer?*

Let pupils suggest ways and then try them. One is to have the box move first *across* the straws, then *along* them, so that they do not roll. Or else, the straws may be kept from rolling by taping them down.

Pupils can make other fair tests of rubbing versus rolling with round toothpicks and flat ones, or with whole marbles and marbles that have cracked in half. They may also:

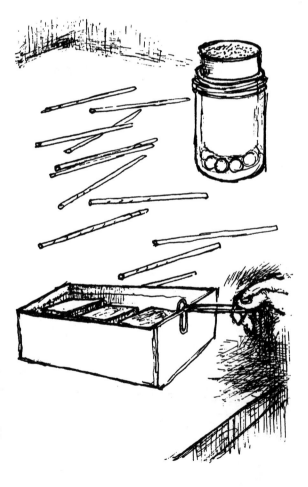

1 Set a roller skate on a playground slide and let it go; then try this with the skate set sidewise, or with its wheels taped to keep them from turning.

2 Fill a tall can with sand, set it in a slightly wider jar, and spin it; then put a circle of marbles in the jar, set the can on them, and spin it.

3 Pull someone on a sled along a concrete walk; then set the sled on round sticks to serve as rollers, and pull it.

4 Watch boxes being unloaded from a truck down an incline having many rollers; then let similar boxes slide downhill without rollers, perhaps on a playground slide.

5 Ask at a garage for an old ball or roller bearing; then clean it and see how it substitutes rolling for rubbing.

PAPER-CLIP CRANKS

Handlebars, wrenches, and doorknobs are just a few of the many everyday devices that enable forces to have great turning effects. A class can readily discover the principle involved in these devices—with nothing more than paper clips.

Let pupils work in pairs, and give each pair a paper clip to straighten out. Then ask the partners each to hold one end of the wire and—without bending it—to make it turn in the other's fingers. *If both squeeze the wire hard, can either one turn it easily?*

Now have the pupils make a right-angle bend in each wire, about 2 centimeters from one end. Then ask one partner of each pair to turn the bent part as a crank, while the other squeezes the wire to keep it from turning. *Who is more successful? If both try to turn the wire in opposite directions, who succeeds?*

Next, ask each partner who is holding the *longer* part of the wire to use *it* as a crank. Also suggest that pupils experiment with cranks of other lengths. *When does a force have a greater turning effect—when it acts on a long crank or on a short one? Does it matter* where *a force acts on a crank—whether close to, or far from, the part of the wire that turns between someone's fingers?*

Finally, let pupils make a second bend in some of the cranks, to form handles. *What effect do these have on how easy it is to turn the cranks?*

These activities show that the turning effect of a force depends on how strong the force is, and also on where it acts. The greater its distance from the center of turning—the *axle* or *pivot*—the greater is its turning effect. This distance, measured at a right angle to the force, is called the *force arm*. A crank is simply a means of providing a long force arm, so that a force has a greater turning effect.

What are some other devices that depend on this principle? Have the class make a list. Also let pupils post pictures of such devices. They may indicate the force applied to each device by an arrow, and the center of turning by a dashed line.

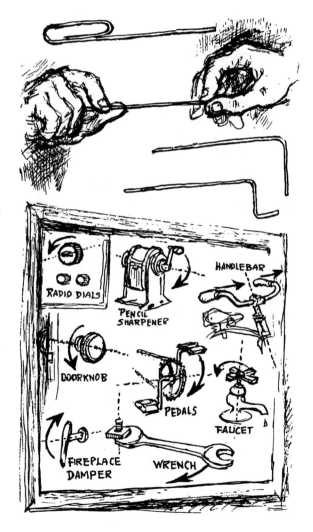

RADIO DIALS

PENCIL SHARPENER

HANDLEBAR

DOORKNOB

PEDALS

FAUCET

FIREPLACE DAMPER

WRENCH

BROOMSTICK PULLEYS

Pulleys are often used to change the direction of pulls—as when a flag is raised by pulling a rope down. They also may multiply pulls—as when a piano is lifted by blocks and tackle. "Pulleys" made from a broomstick can show why this is, and will surprise children by making them seem amazingly strong.

Saw a discarded broomstick in half and give the pieces to two pupils of obviously different weight (because of different mass). Then set two chairs a few feet apart, with their backs toward each other. Have the pupils sit on these chairs, facing each other and holding their sticks by both ends, at arm's length. (**Caution: They should wear gloves to avoid chafing their hands.**)

Now tie a clothesline or other rope to the stick held by the pupil of lesser mass (and weight). Bend it around the center of the other stick, and bring it over one shoulder of the smaller pupil. Then ask a third pupil to pull on the rope, while the other two hold their sticks tightly and lift their feet off the floor.

Whose chair slides? Is this what one would expect with two persons of different mass (and weight)? If these persons exchange places, whose chair slides when the rope is pulled?

Why is this? Can it be due to the difference in mass (and weight) of the two persons? Or is it because the rope pulls singly *on one stick, but* doubly *on the other?*

To find out, also bend the rope around the stick it is tied to. Now it can pull *triply* on this stick, but only *doubly* on the other. *Which chair should slide when the rope is pulled?* Let pupils check their prediction.

When someone pulls on a rope with a force of, say, 10 pounds, this pull acts all along the rope. But if the rope goes around a stick, it pulls *doubly* on the stick. And so, its total pull on the stick is 20 pounds—reduced somewhat by friction. With real pulleys, however, this friction is less, since they turn with the rope.

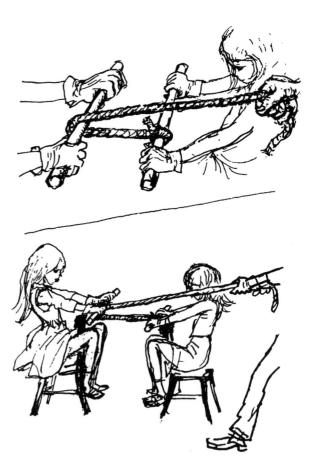

PULLS IN PAIRS

A basic principle of science is that a pull on something is always accompanied by an equal pull, in the opposite direction, on something else. There is a *pair* of pulls. This principle, often misunderstood, can easily be made clearer.

First, get two identical shoeboxes. Then cut a long rubber band to make a single strip of rubber. Tape one end securely to a short edge of one shoebox. Tie the other end to a staple from a paper stapler.

Now set the boxes on a smooth floor or table, end to end. Connect them with the rubber strip, using the staple as a hook, and pull them apart so that the rubber stretches. Then let them go *at the same instant. What happens? Why is this?*

Have pupils try this, too, and then measure how far the boxes move. *How do the distances compare? What does this suggest about how hard the rubber strip pulls on each box?*

Is this also true of pushes? Do they, too, act in pairs? Let pupils find out. Ask a custodian to cut two strips of steel, each 25 centimeters long, from a band around a packing case. Bend one to form a U, and hold it in this shape with a loop of thread. Then, with the U upside down, tape one of its arms securely to the end of a shoebox. Tape the other steel strip to an identical shoebox, to keep the two alike.

Place the boxes in line, with the bent strip between them. Then snip the thread. Let pupils try this, too, and again measure how far the boxes move. *Are they affected equally? Does one seem to be pushed any harder than the other?*

A shoebox, as used here, may be thought of as a person, and a rubber or steel strip as an arm. *And so, if someone pushes or pulls, with one arm, on another person, is there only a* single *push or pull? Or is there a* pair *of pushes or pulls acting on both persons, equally hard?*

Pupils, in pairs, can check this—but they must be able to move easily. They may sit on adjacent playground swings, on boards on wooden rollers, or on wagons, sleds, or skateboards. Or they may, with supervision, float in two inflated boats.

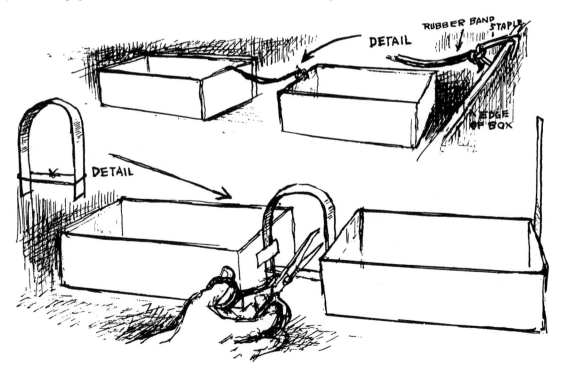

MASSES AND MOTIONS

The concept of *mass,* the quantity of "stuff" in something—not its weight—is a most important one. To help in developing this concept, pupils can investigate how readily things can be moved.

First, let a stretched strip of rubber pull together two identical shoeboxes, or box covers, as in "Pulls in Pairs." Then stick a large lump of modeling clay firmly to the bottom of one. *Now which is moved more readily?* Let all pupils try this, and also note the effect of adding still more clay.

Do nails or stones taped firmly to a box also add to its mass? Will they affect how readily the box can be made to move? Pupils can check this and also whether the effect is the same when two boxes are pushed, as by a bent steel strip. To reduce friction, both boxes may be set on drinking straws, for rollers. (See "Rubbing versus Rolling," page 120.)

When the two boxes are "empty," a stretched rubber strip pulls equally hard on both of them.

Likewise, a bent steel strip pushes equally hard on both. *Is there any reason to think that both pulls, or both pushes, are no longer equal when one box has mass added to it? If not, what would cause the difference in the motions of the boxes?*

Now suppose that two persons of different mass sit on swings, wagons, or skateboards, and gently pull or push on each other with a rope or stick. *Who will move more readily?* Pairs of pupils, of considerably different mass, can check this.

Later, let pupils put clay in two boxes, to add greatly to their mass. *Then can a rubber strip pull the boxes together? What if one box—not both—is set on drinking-straw rollers to reduce friction? What causes this to happen?*

The reason is that friction holds the other box and the table together, making them act as one object, with combined mass. In fact, since the table rests firmly on the floor, and the building is set in the ground, the clay, box, table, building, and the earth act as one object! *How readily can this much mass be moved?*

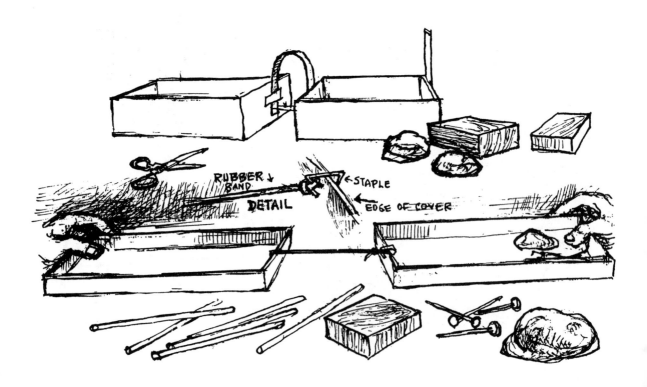

REACTION CARTS

As a sequel to ''Masses and Motions,'' let pupils watch, and then test, carts that illustrate how rockets are propelled. Like rockets, these carts move forward as other things are pushed backward.

First, get a shoebox cover and cut two slits, 5 centimeters apart, in one of its short edges. Slip a rubber band into them and stretch it toward the opposite end of the cover. Hold it with a thread fastened to the opposite end. Then rest a slender stick on the edge of the cover between the slits, with one end against the stretched rubber band.

Now lay out many drinking straws, for rollers, on a table or smooth floor. Set the cover on them and **ask everyone to stand back.** Then snip the thread with scissors, or burn it with a match. *What happens to the stick? To the cart?*

Once it is released, on which two objects does the stretched rubber band act? How do its forces on these objects compare? How would the motions of the objects compare if they had equal mass and were retarded equally by friction?

Next, get a lightweight paper box and cut a 1-centimeter hole in one end. Blow up a fairly large balloon and hold it by the neck. Put it in the box, with its mouth through the hole and only the lip on the outside. Then set the box on drinking straws for rollers and let go. *What happens? Why? What two forces and what two things are involved?*

The propulsion of these carts may be said to be a *reaction* to the *action* of the stick or of air. Both the action and the accompanying reaction result from a pair of equal forces exerted by the stretched rubber band or balloon, on two different masses. In each case, one of these masses is the cart. *What is the other?*

Other common illustrations of this principle include:

The retreat of a rowboat when someone steps from it to a dock

The rotation of a lawn sprinkler when water pressure pushes jets of water in one direction and a rotating pipe in the other

The forward propulsion of a jet airplane when gases produced by the burning fuel shoot backward

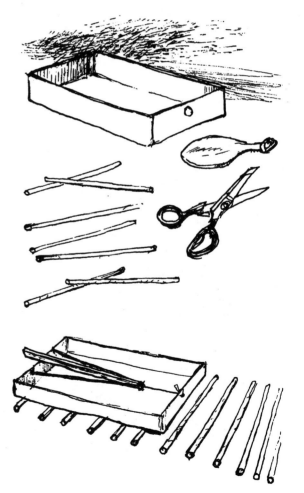

WHIRLED WATER

Many children have seen someone swing a pail of water around without having it spill. Some have also seen a spin drier, in which wet clothes are whirled around in a drum that has holes in it. The same principle—often misunderstood—is involved in both cases.

In presenting this principle, a good way to start is with a few paper cups and long pieces of string. Have pupils poke two small nail holes below the rim of each cup, opposite one another, and put the ends of a string through the holes and knot them. Then, preferably outdoors, let the pupils put water in the cups and swing them around. *What does the water do?*

Now have the pupils poke a small hole in the bottom of each cup, put in some water and swing it around. *While being swung around, does water come out of the hole even when the cup is upside down?*

Next, let the pupils poke many holes in the bottom and side of each cup, from the inside, put in a very wet rag and swing it around. *What keeps the rag from going off with the water?*

Pupils may also hang up these cups by the strings, twist the strings around many times, and put a very wet rag in each cup. *What happens when a cup is let go and whirls around fast? How is this like what happens in a spin drier?*

Another way to show this is as follows: Wrap a narrow strip of cloth around the rim of one wheel of a roller skate. Hold it in place with a rubber band, and wet it. Lay the skate on its side on a sheet of colored paper, with this wheel down. Then spin the wheel fast. *What do the drops of water do? What are their paths like?*

It is often said that "centrifugal force" keeps water from spilling out of a pail or cup that is being swung around. It is also said that water is "pulled out" of wet clothes in a spin drier. Unfortunately, both these statements are false.

Pupils, however, can be helped to understand the correct concept by letting them take part in this series of activities:

1 Roll a marble across a smooth, level table. *What is its path like?* All moving things, including bowling balls, hockey pucks, and water drops, *tend* to move like this. They go straight unless they are pushed or pulled sideways, out of a straight line.

2 Blow on the marble from the side as it rolls along. *How is its path changed by the push of the air? Once this push stops, does the marble keep changing its direction?*

3 Ask someone to hold a strip of cardboard on edge on the table, and curve it gently. Then roll a marble along the inside of the curve. *What pushes sideways against it, and prevents it from going straight? If the cardboard is quickly lifted straight up as the marble rolls along it, what does the marble do?*

4 Put a marble in a round cardboard box cover, and make it roll around and around. *Now what pushes sideways against the marble? What would the marble do if it were not pushed sideways?* To test this, invert the box cover on the table, over the marble, and make the marble roll around inside it. Then lift up the cover. *What does the marble do?*

5 Cut a few gaps in the rim of the box cover, big enough for marbles to fit through. Then put a few marbles in the cover, and make them roll around. *When a marble happens to be at one of the gaps, can the rim of the cover push against it? Then what does the marble do?*

The above activities show that marbles, once moving, tend to keep moving—straight, in whatever direction they are going at the time. *No* force is needed to make them do this. But a force is needed to *prevent* them from going straight!

Water acts in the same way. When it is swung around in a cup or pail, the container keeps pushing it out of a straight line. The water tends to go straight, but the container keeps pushing it toward the center of the circle and makes it change its direction continually. And so it goes in a circular path, with the container, and cannot escape.

However, if there are holes in the bottom or side of the container, water is able to escape. Then the container cannot push the water out of a straight line, and it simply keeps on going.

Similarly, in a spin drier, water is not *pulled out* of the wet clothes. Instead, it escapes through the holes in the drum, and just keeps going. Meanwhile, the drum pushes against the clothes, toward the center of the circle, and prevents them from going off with the water. And so the clothes are *pushed away* from the water!

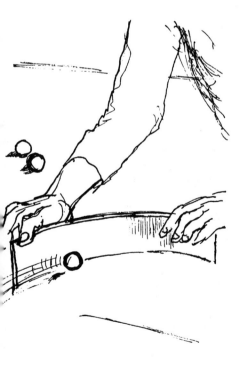

PATHS OF MOTION

Really understanding how things move involves being able to visualize the direction, and changes in the direction, of motion. Pupils can develop this skill through a game that combines learning with fun.

Start by drawing a line on the chalkboard, like A, B, or C. Then let the class try to think of some object that moves along a path like this. *Might it be a marble rolling across a smooth floor? A ball falling freely? Or an airplane landing? Could this line also represent the motion of something else?*

Once pupils get the idea, draw the path of another moving object and see who can name it. Then draw other paths of motion, one at a time, of gradually increasing complexity. Let pupils who answer correctly draw the paths of moving objects that they think of, for their classmates to try to name.

Another approach is to let the class draw the paths of motion of objects suggested by the pupils, such as:

The taillight of a motorcylce as it turns a corner

The handle of a pencil sharpener as a pencil is being sharpened

A baseball that is pitched, and then batted into the stands

A hockey puck during a hockey game on ice (like D)

A ball bouncing on the floor, or down a flight of stairs (like E)

A marble rolling back and forth inside a shallow bowl

A washer swinging on a thread inside a moving elevator (like F)

An insect crawling toward the center of a spinning record

A pebble stuck in the tire of a moving automobile

A reflector on the pedal of a bicycle that someone is riding

Close observation will help, and disputes can be settled by experimentation. For example, to see what kind of path a pebble in a tire makes, tape a piece of chalk to the side of a can. Then roll the can along a chalk tray so that the chalk rubs against the chalkboard. Or, to record the motion of a bicycle pedal, fasten a flashlight to it. Then take a "time" picture at night.

MASS IN A MAZE

This game depends on a basic principle of science. In playing it, pupils learn about inertia—in a sense with their muscles as well as with their minds!

To make the game, get a box cover, perhaps from a shoebox, and a marble or a metal nut about ½ inch across. Then construct a maze inside the cover by pasting in narrow strips of thick cardboard. The maze may have any pattern, from simple to complex.

Now, before pupils play the game, show them this: Put a paper clip in a paper cup lying on its side. Then move the cup fast, open end first, and stop it suddenly. *What does the paper clip do?* Next, with the paper cup still on its side, put the paper clip back in. Then move the cup suddenly, bottom first. *This time what does the paper clip do?*

Let pupils try these things—and also test other objects, such as a coin in a box or a block on a sheet of paper. They will find that any object they test, once moving, tends to keep moving. And, when not moving, it tends to stay still.

In short, things tend to keep doing whatever they have been doing, whether moving or standing still. This tendency is called *inertia*. Inertia is characteristic of all mass—even that of people. For example, when an automobile starts to move, the people in it tend to stay still. And when it stops, they tend to keep moving!

Now pupils are ready to play the game. The box cover is laid on a table, and the marble or nut is put at "start." Then each player, *without lifting or tilting the cover,* slides it this way and that until the marble or nut reaches "finish."

There are two ways of doing this—both depending on inertia. One is to slide the cover to get the marble or nut moving, and then to stop the cover abruptly, letting the marble or nut keep moving. The other is to slide the cover abruptly, letting the marble or nut stay still.

Records may be kept of how long each player takes (see "Swinging Second-Timer," page 19). Then a graph of successive times will show how much a player's skill improves with practice.

Vibrations and Sounds

Drums, whistles, and horns all attest to the fact that sounds are intrinsically appealing to children. It is great fun for them to produce sounds in various ways, and this can lead to their learning much about music as well as other sounds. They can also learn a great deal by investigating the motions of objects that swing, bounce, or sway—motions like the vibrations that produce sounds, only slower.

Pupils should become aware that many familiar objects vibrate—at different rates and with different degrees of vigor. Through firsthand experiences they should discover that these vibrations can often be controlled, and that when they are rapid enough, sounds result. From this should come an understanding of the principles employed in musical instruments, so often used for making the air vibrate.

Youngsters are aware of a great variety of sounds. This is clearly shown by their vocabu-lary—with words like *boom, clang, hiss, moan, pop, roar, squeak,* and *whir*—and by their imitative noises, such as *ack-ack-ack* and *brr-brr-brrm.* Older pupils can appreciate the importance of sounds as communication, as music, and as noise. They can experience the effects of sounds on work, study, and relaxation. And they can begin to realize the need for the control of sound pollution in an increasingly noisy world.

Investigations of vibrations and sounds are, in themselves, interesting and challenging. But they also contribute to more general objectives. These include competence in carrying out experiments and in making observations, proficiency in measuring and in keeping records, and facility in working cooperatively with others. Above all, perhaps, is a sense of accomplishment that can come, for example, from making a musical instrument and playing a song!

SOME IMPORTANT OBJECTIVES

Attitudes and Appreciations to Be Encouraged

Vibrations are important in our lives, to a large extent because they are the cause of sounds.

The more we understand about vibrations and sounds, the better we can use and control them.

Scientists learn about vibrations and sounds by conducting experiments, making observations and measurements, noting relationships, and testing ideas and hunches.

The nature and rate of vibrations, and the loudness and pitch of sounds, depend upon definite and measurable factors, not on mere chance.

It is fun to play songs, even if somewhat out of tune, on a simple musical instrument that one has made.

As people live and work more closely together and make greater use of motors and machines, the environment becomes increasingly noisy.

Music and noise, including cheering and applause, have an effect on our state of mind and on how well we work or play.

To protect our hearing, and for the sake of our mental well-being, it is important to reduce the loudness of noise and other undesirable sounds.

Because not all people enjoy the same sounds, no one has the right to inflict sounds on others against their will.

The enjoyment of music is important, even though it may not lend itself to scientific study and measurement.

Skills and Habits to Be Developed

Being aware of the variety of commonplace vibrations and sounds, and being able to describe the differences among them

Constructing simple musical instruments using stretched strings, air columns, or pieces of wood, and playing scales and songs on them

Experimenting with vibrating objects to discover the effect of changing various factors such as size, length, tautness, kind of material, and mass.

Predicting the effects that changes made in objects will have on their vibration rates and the pitch of the sounds they may produce

Making sounds louder by using sounding boards and other means

Determining the rates of vibration of objects—such as metal or wood strips, pendulums, and the wings of insects—by measurement or by comparison with sounds of known vibration rates

Identifying the objects that vibrate in various musical instruments, and in the production of the human voice, insect songs, and other common sounds

Comparing the speed and clarity with which sounds travel through air and other materials

Detecting echoes, and relating their delay to the distance of the surfaces which reflect the sounds

Measuring the speed of sound in air by timing the delay of echoes with a pendulum

Graphing the relationship between two quantities—for example, the rate of vibration of a steel strip or of a pendulum, and its length

Reading, with understanding, such terms as *air column, echo, frequency, octave, pitch, rate, sounding board, tautness, vibration, xylophone*

Facts and Principles to Be Taught

Many objects vibrate, including drums, buildings, telephone poles and wires, bodies of water, and even the ground.

Objects vibrate in various ways and at widely differing rates.

To produce sounds that we are able to hear, vibrations must occur very rapidly.

Most sounds that we hear are ordinarily transmitted by air that is set into vibration by various vibrating objects.

Sounds differ in pitch, and this depends upon the rate of vibration, or frequency.

One way to make sounds louder is to set larger surfaces, such as the wood of a violin or piano, into vibration.

Many creatures besides people communicate by means of sounds, but often we do not notice or cannot hear these sounds.

Sounds travel faster and better through some materials, such as taut strings, than through others, such as air.

Sounds travel at a high, but measurable, speed in air—although much slower than light.

Echoes are sounds that are reflected by distant objects; the farther the reflecting object, the longer an echo takes to return.

SOUND RACE

Can-and-string telephones are well-known toys which show that taut strings conduct sound well. Besides this, however, they can be used to compare the speed of sound in string, or even in wire, and in air.

First let pupils make such telephones. Have a pair put the ends of a 20-meter or longer string through nail holes in the bottoms of two cans, and knot each on the inside. Next, let them walk apart with the cans until the string is taut.

Then, when one pupil speaks into the ''mouthpiece,'' the sound can be heard by the other pupil, who holds the ''receiver'' to one ear. *Can tapping or scratching the bottom of one be heard at the other end? The ticking of a clock?*

Now, to compare the speed of sound in string and in air, one pupil should hold a can to one ear, while the other pupil taps the other can sharply with a large nail. The first pupil will then hear *two* sounds, one with each ear. *Which is heard first? And so, does sound travel faster in air or in the string? Might it travel faster, or slower, in wire?*

TUNED TUBES

Cardboard tubes from inside rolls of paper towels, waxed paper, and aluminum foil can easily be tuned to play songs. By doing this, a class can learn some principles on which wind instruments depend.

Ask pupils to bring such tubes from home. Then show them how to produce sounds by holding the tubes loosely and tapping them lightly. Also demonstrate how to "play" the tubes by blowing across their ends. *Do they all sound alike? If not, what might cause this?*

Next, ask a few pupils with similar tubes to "play" them. Then cut off about 1 centimeter from the end of one tube, 2 centimeters from another, and so on. *What does this do to the* pitch?

Now let pupils shorten their tubes, a little at a time, or make them longer by taping on extra pieces, until the sounds match a scale played on a piano. With such tuned tubes, eight pupils—or the whole class—can give a concert!

Pupils may also carry on these investigations:

1 Hold two similar tubes together, end to end, and blow across the end of one. *How does the pitch compare with that of a single tube? The length of the* air column *with that in a single tube?*

2 Find one tube that just fits inside another, so that you can slide it in and out. *How does doing this affect the pitch? Why is this? What band instrument has its pitch changed by sliding a tube?*

3 While blowing across one end of a tube, close the other end with your palm. *How does the pitch change? Is the air column changed? If so, how?*

4 Let water from a faucet fall into a tall tumbler or jar, and listen closely. *What happens to the pitch? Is this because of more water in the vessel, or less air? How might you find out for sure?*

5 Get eight similar bottles and, with seven friends, blow across their tops. *How can you tune the bottles to play a scale, and then some songs?*

MUSICAL DESKS

Violins, harps, pianos, and many other musical instruments produce sound by means of tightly stretched strings or wires. These vibrate rapidly when bowed, plucked, or struck.

Such strings are tuned by changing how taut they are. Also, in some stringed instruments, the players further control the pitch of the strings by changing the length of the portions that vibrate. They usually do this with their fingers.

To make all this more meaningful, let pupils have firsthand experiences. Give each one a length of *strong,* thin string or thick thread—to go around the top of a desk or table, over and under. Help to tie it tightly with a *square knot* (see illustration). Then have the pupil make it more taut by slipping two pieces of scrap wood beneath it.

When plucked, the raised portion of the string or thread should produce a clear tone. Its pitch can be changed by moving the supports. *What is the effect of changing the tautness? Of changing the length of the portion that vibrates?*

Some string and thread, and also monofilament fishline, hold their pitch rather well. With eight such strings, properly tuned, a pupil can make music. Or eight pupils can make their desks into one-note instruments and play songs by plucking the strings as a "conductor" points to each one in the right order. Sometimes the quality of the tone is improved by removing books, lunches, and baseball gloves from inside the instruments! (See "Louder Sounds," page 136.)

It is also instructive for pupils to:

1 Tap the strings lightly with a pencil or small stick.

2 Stroke a string with a violin bow, fairly close to one support.

3 Play a string while lifting up one of the supports slightly.

4 Pinch a string firmly between two fingers while playing it.

5 Hold an ear to the desk or table while playing a string.

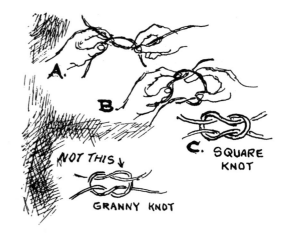

SCRAP-WOOD MUSIC

With scraps of wood, a class can make simple xylophones and thus learn about some factors that determine pitch—as well as make music! (The word *xylophone,* from the Greek, means "wood sound.")

Cedar shingles, possibly scraps from new housing, work well. First split them into strips 2 to 3 centimeters wide. Then let the pupils saw various lengths, 10 to 20 centimeters long, from the thicker ends. Other wood may be used, too, such as narrow strips from crates or from a lumber yard. (Pupils can saw safely with coping saws. Show how to do this properly, on an old table or large wooden box.)

When several pupils have cut lengths of wood, let them stand in line on a hard tile or concrete floor, in order of the length of the pieces. Then ask them to drop each piece in turn, starting with the longest. *What do you notice, in general, about the pitch? If there are exceptions, what might cause these?* Suggest that the pupils compare the width and thickness of the pieces, as well as the relative mass or weight of the wood. Also let them split some of the pieces further, or saw them to shorter lengths, and test their pitch again.

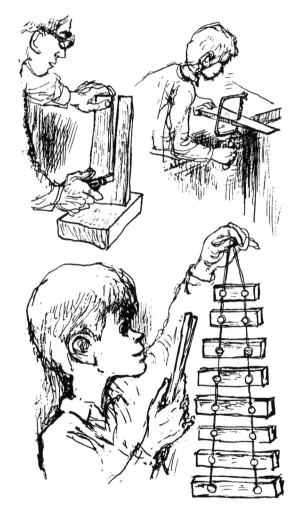

At another time, perhaps, play a scale slowly on a piano, or even on a harmonica. After each note have pupils drop their pieces of wood until they hear one of matching pitch. When they have eight pieces that sound a scale, they may hang these up on a string to make a xylophone. They should drive two thumbtacks, with the string wrapped around each, into each piece of wood, about one quarter of the way in from its ends. Then they can play songs by tapping the pieces with a stick.

In making scrap-wood music, be sure that everyone contributes something—even if a bit out of tune. It takes eight or more pieces of wood to play a conventional scale, but four may sound a pleasant chord, such as C-E-G-C (do-mi-sol-do).

LOUDER SOUNDS

Pupils can observe that the sound from a plucked string across a desk (see "Musical Desks," page 134) is not so loud as that from a plucked string on a guitar. One reason for this is that the shape and material of the guitar's body makes the sound louder. Most stringed instruments have a kind of *sounding board* that helps to make the sound louder. Children can investigate some of the objects and materials in their classroom to see which make louder sounds than others, or which make the best sounding boards. With their hands, voices, and ears they can also learn other ways to make faint sounds louder.

Ask each child to bring a pocket comb to school. Then show the class how to hold a comb in one hand and stroke the teeth with the thumb of the other hand. Let each child try this with his or her own comb. Next, let the children compare the sound from a comb stroked in the hand with the sound produced when one end of the comb is held tight against some surface such as a desk top. *From which is the sound louder—comb in hand, or comb against desk top?*

Now divide the class into several small groups, each group with a list of various surfaces or objects within the room—teacher's desk, bulletin board, window, door, chalkboard, floor, bookshelf, sink, etc. Add some blanks for pupils' ideas, too. Then ask each group to investigate how loud a sound can be made by stroking a comb while pressed against each surface. *Which makes the best sounding board of all? Which surface produces the faintest sound? How do surfaces made of wood compare with surfaces made of other materials? How does pressing a hand against a good sounding board change its effectiveness?*

At another time ask a pupil to go to the front of the room and read aloud or talk to the class in a low voice. Let the rest listen, with their hands cupped behind their ears. Then ask them to cup their hands in the opposite direction, in front of their ears. *When is the sound louder?* Now ask the reader to read from the back of the room, keeping his or her voice the same as before. *How should the listeners, still facing front, hold their hands to hear best?*

BUZZ RATES

Pupils may think that piano strings are used only to produce musical notes, but they can also be used to tell how fast *other* things vibrate. For example, with the help of a piano, a pupil can tell how rapidly a mosquito's wings beat as it flies.

The pitch of a sound made by a vibrating object depends on the rate of vibration (or *frequency*). Objects that produce the same pitch vibrate at the same rate. On a properly tuned piano, the strings are adjusted to vibrate this many times each second:

middle	C	262	F#	370
	C#	277	G	392
	D	294	G#	415
	D#	311	A	440
	E	330	A#	466
	F	349	B	494

For each octave above one of these notes, the vibration rate is double; for each octave below, it is one-half. C above middle C, for example, has twice the rate of middle C, or about 524 vibrations per second. *What would it be for a note an octave above G?*

Ask your pupils to write down how rapidly they think a mosquito's wings vibrate as it flies. Then let each pupil find out, following these directions:

Catch a mosquito or other small flying insect and put it in a jar covered by a piece of cheesecloth or a handkerchief. Copy the list of vibration rates shown above. Then take both the list and the jar to a piano. Put your ear to the jar and listen to the pitch of the insect's buzzing. Then pick out the note on the piano that most closely matches it. *According to the list, how rapid are the wingbeats of the mosquito? How rapid are the wingbeats of a housefly? A wasp? How might you find out how fast the diaphragm of your car horn vibrates?*

When you sing, your vocal cords vibrate as air from your lungs moves past them. *What is the fastest that you can make them vibrate? How rapidly does the air vibrate when you whistle? What is the frequency of the highest note on a harmonica?*

UNHEARD VIBRATIONS

On every hand pupils can see objects vibrating—back and forth, up and down, or side to side. Such vibrations may be too slow to produce sounds. Even so, some of them are useful, such as those of clock pendulums, diving boards, and pogo sticks. Others are unpleasant or dangerous—as when ships roll, suspension bridges swing, or buildings sway during earthquakes.

Groups of pupils can easily investigate vibrations like these with the help of pieces of steel bands from around packing cases or lumber. Ask a custodian to save some, and to cut several straight pieces about 2 feet long.

Each group should rest one end of a strip on a desk, and weight it down with blocks or books set even with the edge of the desk top. Then someone should lift up the free end of the strip slightly, and let go. *How many complete up-and-down vibrations does it make in 1 minute—or would it make if it kept vibrating that long?* Have pupils check this and keep a record.

Do the vibrations gradually slow down? Pupils can compare the number of vibrations during three 10-second periods—at first, later on, and still later. *Could a steel strip be used to measure time accurately?* (See "Swinging Second-Timer," page 19.)

How can you change the rate of vibration? Pupils can move a strip so that more, or less, of it is free to vibrate. The fixed part should be clamped tightly by pressing a block of wood down on it, exactly even with the edge of the desk top.

As pupils change the length of the part that vibrates and find the vibration rate of each length, they should keep records. Then they may make a bar graph to show the relationship. They should mark a scale of the number of vibrations per minute along a horizontal line, and show the various lengths of the vibrating part by vertical bars.

As the vibrating part is made shorter, do its vibrations soon become too rapid to count? Do they eventually produce a hum or buzz? If so, how can you find the rate of vibration? (See "Buzz Rates," page 137.)

Later, groups may carry out related investigations, such as these:

1 *Does the rate of vibration of other objects also depend on the length of the part that vibrates?* Try yardsticks, thin rulers, plastic spoons, hacksaw blades, and straightened-out bobby pins. *Can you make these objects vibrate rapidly enough to produce sounds?*

2 Tie the ends of a length of shade cord or strong twine, about 30 feet long, to two chairs. Stand them as far apart as possible, and have someone sit on each to keep the cord or twine taut. Then pluck it at the center, sideways, and note its vibration. *What is the effect of making it even more taut? Can you make it taut enough to produce sound?* (See "Musical Desks," page 134.)

3 Set a can on a windowsill and fill it to the brim with water. Stand so that you see a bright

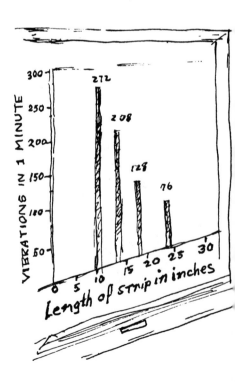

reflection in the water. Then touch the water surface with a pencil point and observe the ripples. *What happens when you move the pencil point up and down as fast as you can?*

Now tap the side of the can. Listen to the sound, and observe the tiny ripples. *How must the vibrations of the can compare in rate with those of your pencil point?*

4 Make a "chain" of a few thin rubber bands, or use a slender spring. Hang one end on a support, and hang a paper cup on the other. Put some paper clips or nails in the cup, pull it down a little way, and let go. *What happens?*

How does the amount of stuff—mass—in the cup affect how rapidly it bobs up and down? Count the number of up-and-down vibrations in 10 seconds. Do this for different masses, and make a graph. *What does this graph show? Could you use this method for measuring mass?*

5 Get a straight and uniform lattice strip of thin wood, 12 feet long, from a lumber yard. Hold it in the middle so that it balances, and make its ends spring up and down—at the "natural" rate. *How many vibrations does the strip make in 1 minute?* Check this, and keep a record. Ideally, several pupils should do this and compare their findings.

Next, shorten the strip by sawing off a 1-foot piece, and again find the natural rate of vibration. Saw off another foot, record the new rate, and so on—until the vibrations are too fast to count. Then make a graph to show the change in vibration rate with length.

From this graph, what do you predict the vibration rate to be of a piece only 6 inches long? To check, cut off a 6-inch piece and drop it on a hard floor. Match its pitch on a piano, and then find its vibration rate. (See "Buzz Rates," page 137.)

SOUND BOUNCE

It is well known that an echo is caused by the reflection of sound by a building, cliff, or other large object. The sound bounces off the object—somewhat as a rubber ball rebounds when it hits a wall.

Sharp, distinct echoes can help a class to appreciate how fast sound travels. For this purpose a large, plain outside wall of a building works well. It should face a wide open area, and there should be no other large objects nearby.

Have the class stand 100 meters or so from the wall, at a time when there is little noise and no wind. Ask one pupil to bang the bottom of a large can with a stick, while the others listen for the echo from the wall. Then walk toward the wall together, and let the pupils repeat this at various distances from it. *Do you notice any change in how quickly the echoes return? If so, what causes this change?*

By using a pendulum (see "Swinging Second-Timer," page 19), pupils can find the approximate speed of sound. They should stand about 100 meters from the wall. Someone should bang the can very regularly, in time with the swinging of the pendulum. Each bang should occur at the precise instant the pendulum is at one end of its swing. *Then do the echoes return exactly when it reaches the opposite end of its swing?*

If not, the pupils should move closer to or farther from the wall, and again bang the can each time the pendulum is at one end of its swing. At some distance from the wall the echoes will return exactly when the pendulum reaches the other end of its swing. Then the sound will travel to the wall and back in the time the pendulum takes to make one-half of a complete swing.

How long does the pendulum take to make half a swing? And how far is it from the can to the wall and back? (See "Wheel Measure," page 18.) *Then how far does the sound travel in 1 second?*

PENDULUM PREDICTIONS

A fascinating project for a class is to discover how the length of a pendulum affects its swinging or vibration rate. Pendulums, themselves, can show this as a graph—from which predictions can be made.

Give out metal washers or nuts, straight pins, and pieces of thread of various lengths from 15 to 150 centimeters. Let each pupil make a pendulum and count how many times it swings in 1 minute. (See "Swinging Second-Timer," page 19.) Then have the pupil check this rate and record it.

Next, make 21 evenly spaced marks along the top of a chalkboard or bulletin board. Number them by 5s, from 0 to 100. Then let each pupil hang up his or her pendulum at the position that corresponds to the number of swings per minute. The loop tied at the upper end of each pendulum's thread may be taped to a chalkboard or hung on a pin stuck in a bulletin board. All the loops should be at the same level.

Now draw a smooth line on a chalkboard, or tack a string to a bulletin board, to show the gen-

eral change in length of the pendulums. *Are any of the washers or nuts not at all close to this line? If so, why may this be? Are there any big gaps?* Let pupils suggest how the line of the graph may be made smoother and more accurate, and let them carry out their suggestions.

On the basis of this curve or graph, pupils can make, and record, predictions about other pendulums, and then test them. Thus:

1 *How short would the fastest-swinging pendulum be whose swings can still be counted?*

2 *What would the rate be of the longest pendulum that can be hung up inside the room?*

3 *What would the rate be of a very long pendulum—perhaps one hanging out of a window?*

4 *Would the weight, or mass, of the swinging object affect a pendulum's rate? For example, would a pendulum made with two or three washers fit on the curve of one-washer pendulums?*

5 *Would swinging through a bigger or smaller distance affect a pendulum's rate? Or would its rate stay the same even while the swinging is "dying down"?*

RESONANCE IN RUBBER

Resonance sometimes results when one object or material "feels" the vibrations produced and sustained by another. Some objects vibrate easily at a particular frequency or rate produced by a separate vibration maker. When those objects also vibrate, they are said to resonate. One example is the "singing" of a thin glass tumbler when a moist finger is rubbed around its rim. Another is the sound of a violin string when a bow is moved across it at just the right pressure and speed.

Children can feel what resonance is like by lightly pushing a nearly vertical piece of chalk across the board. The chalk should be slanted slightly so that the point of the chalk on the board is pushed, not dragged. When the pressure and the angle are right, the chalk will "stutter," leaving a series of evenly spaced dots. The child can feel this stutter, which occurs much too rapidly for the child to have made the dots on purpose.

Another example of resonance is the sound-caused patterns of salt on a tightly stretched rubber diaphragm. In a place and at a time when singing and laughter will not disrupt neighboring classes, let your own class experiment as follows.

Cut the neck from a large, round rubber balloon. Stretch the remainder evenly and tightly across the opening of a large juice can, holding it in place with rubber bands. Then sprinkle some table salt across the top of the rubber diaphragm.

Next, roll a sheet of construction paper into an open cone, and tape it to make a megaphone. Now, one pupil should direct the megaphone at the salt and loudly sing "Ah," varying the pitch slowly until the salt is observed to dance on the rubber. When the salt begins to dance, it is evidence that portions of the rubber are resonating.

If the pitch is kept constant when the salt begins to dance, the salt will move into patterns outlining cells. The cells are formed by portions of the rubber diaphragm bouncing up and down (resonating) in accord with the vibrations that produced the sound. *Can the same pattern be produced by another person singing at the same frequency? What if the person who produced the first pattern sings a* different *note?* Suppose that everything is kept the same, except that paprika powder is used instead of salt. *Will the pattern be the same for the same frequency? What happens to the resonance patterns if the tension on the rubber is changed?*

Suppose that for some reason the rubber diaphragm had to be removed. *Could resonance patterns be used to make sure that the rubber was replaced in exactly the same way it was before removal?*

Magnetism and Electricity

Daily life is truly dependent upon magnetism and electricity. Through them we benefit from tremendous amounts of energy, a great deal of enjoyment, and a degree of convenience too often taken for granted. Youngsters encounter applications of them on every hand—in toys, lights, stoves, motor-driven tools, television sets, and numerous other devices. And when these children become adults, practically all of them will rely on magnetism and electricity, directly or indirectly, for their livelihood.

It seems obvious, therefore, that all pupils have a right to learn the basic concepts of magnetism and electricity. The subject can be made highly interesting to them, creating enthusiasm for further study which, for some, may lead to careers. Moreover, it can serve as a medium for providing other aspects of general education, including oral and written communication, mathematics, and manual skills.

Much of the fascination of magnetism and electricity stems from their strangeness. One example is that they can cause pushes and pulls on things at a distance, often without any discernible material connection. Their curious behavior has long prompted scientists to wonder about their fundamental nature. And the resulting investigations have led to some of the most basic science concepts involving the structure of matter, the bases of chemical change, and the nature of energy.

It is essential that children learn to respect, yet not fear, electricity. They must, of course, be taught the hazards of lightning, fallen wires, and worn appliance cords, and the danger of poking things into sockets and outlets. At the same time, they should be helped to feel confident that they can control and use electric currents safely. And they should be encouraged to experiment freely with magnets, flashlight cells, and the electric charges that they can produce by rubbing.

SOME IMPORTANT OBJECTIVES

Attitudes and Appreciations to Be Encouraged

Electric outlets and sockets in homes and schools are too dangerous for beginners to use in investigations, but one or a few flashlight cells ("batteries") are safe.

It is hazardous to fly kites near electric transmission lines, climb the poles or towers, go near wires that have fallen, or throw stones at the insulators.

One should seek shelter from lightning—preferably inside a building or an automobile—and especially avoid being out in the open, under a tree, out in a small boat, or in swimming.

Electricity may do great harm and therefore must be treated with understanding and respect, but it should not be feared unreasonably.

Magnetism and electricity are fascinating to experiment with and learn about, and an interest in them may lead to a life's work.

A person often is able to carry out good scientific research—for example, in magnetism—with only commonplace things.

In conducting comparison testing—such as of electromagnets—one must be sure that the tests are fair, with all conditions exactly the same except for the one being tested.

Scientists have learned and are still learning basic principles of magnetism and electricity by doing experiments, making measurements, and testing possible explanations of their observations.

Although magnetism and electricity can act on distant things through empty space and seem to act mysteriously in other ways, too, there is no reason to consider them to be supernatural.

Even though no one completely understands the nature of magnetism and electricity, we are able to control them and put them to use in countless ways.

In the modern world, electricity and magnetism provide tremendously important means of transmitting energy and using it to do work.

Except along some transmission lines, electrical energy seems not seriously to pollute the environment, but both nuclear and fuel-burning electrical generating stations do pollute it.

Skills and Habits to Be Developed

Investigating whether various materials are magnetic, and whether magnetism is able to act through them

Magnetizing steel objects by stroking them with magnets, or by winding coils around them and connecting the coils to flashlight cells or batteries of cells

Handling magnets carefully so that they are not weakened by dropping or banging them

Comparing the strength of magnets or electromagnets by accurate measurement

Making a simple magnetic compass, and using it to determine direction

Explaining how to find out whether objects are magnetized, and if so, where their north-seeking and south-seeking poles are located

Connecting a flashlight lamp ("bulb") to a flashlight cell ("battery") so that the lamp lights

Constructing simple lamp sockets, cell holders, and switches, and connecting them properly with wires to make light circuits

Carrying out simple but valid scientific research—for example, on how well various materials conduct electricity

Producing electric charges on various objects, and testing whether these charges are positive or negative

Using correctly, in speaking and writing, such terms as *battery, cell, electromagnet, insulation, lamp, magnetic north, north-seeking pole, positive charge, repulsion, static*

Facts and Principles to Be Taught

Magnets attract mostly iron and steel; however, some materials other than these are magnetic, also.

Magnetism is able to act at a distance, through nearly all materials, but it becomes rapidly weaker as the distance increases.

Steel objects are commonly magnetized, but they usually can be made much stronger magnets by stroking them with strong magnets.

An ordinary magnet has two different poles, *north-seeking* and *south-seeking,* named for what they do when the magnet is able to turn freely.

Like magnetic poles tend to repel each other, and *unlike* poles tend to attract; however, a pole of either kind and a piece of unmagnetized iron or steel tend to attract each other.

The earth is a huge magnet and therefore affects a freely turning magnet, causing it to act as a compass—to take a consistent position that indicates *magnetic* north and south.

The North Magnetic Pole of the earth—so-named because it is in the north—is actually a *south-seeking* pole.

When electricity flows through a wire or other piece of metal, the metal acts like a magnet—even if it is not iron or steel.

Electromagnets, consisting of coils wrapped around iron, are used in many important ways—for example, as lifting magnets, in electric motors and telephones, and for generating electric currents.

For an electric lamp ("bulb") to light, it must be part of a continuous circuit so that electricity flows through it.

Electricity can flow readily through some materials, known as *conductors,* but not through others, called *nonconductors* or *insulators.*

A cell, or a *battery* of cells, is used to make electricity flow in a circuit, and a switch is used to open or close a gap in a circuit.

When flashlight cells are used a great deal, as for operating electromagnets, they rapidly lose their energy and become weak.

Electrical energy is tremendously important in modern life for producing motion, heat, light, and sound.

Many materials acquire *electric charges* when they are rubbed with other materials; they are then able to attract light objects, make sparks, and cause fluorescent lamps to flash.

An electric charge on a material is either of two kinds: *positive,* if the material tends to repel wool that has been rubbed with rubber; or *negative,* if it tends to repel the rubber.

Objects with like electric charges tend to repel each other, and objects with unlike charges tend to attract; however, an object with either kind of charge and one without a charge tend to attract each other.

Both conductors and nonconductors may possess electric charges that are *static,* but conductors also permit charges to flow along them as *electric currents.*

MAGNETIC TIGHTROPE

This simple setup will fascinate children, especially if made before they come into the room. When they notice it, their reactions are likely to be most interesting!

To set it up, use a strong magnet—possibly one from a junked radio speaker from a repair shop—and a length of thread tied to a paper clip. Anchor the other end of the thread in some way, perhaps by a book. The paper clip should come close to the magnet, but be prevented from touching it by the thread. Then the pull of the magnet will keep the paper clip suspended.

Give pupils a chance to play with the setup and think about it. *What will happen if a sheet of plastic is slipped between the magnet and the paper clip? A piece of aluminum foil? A pane of glass?* Let pupils make predictions, and then check.

Also let them test a variety of other materials, slipping them between the magnet and the paper clip carefully so as not to push the paper clip away. *What common material causes the paper clip to drop off?* There is only one!

ELECTROMAGNETIC SWING

This toy makes learning fun. It shows that a magnet affects a wire or strip of metal in which there is an electric current, and that electric energy can cause motion—as well as heat.

To make it, find a short scrap piece of a two-by-four, a little longer than a flashlight cell ("battery"). Then cut two L-shaped pieces of aluminum from a disposable pan, and tack these to the wood. Rest on them a swing cut from an aluminum pan, or fashioned of bare copper wire. It must be free to swing, just clearing one end of a strong magnet lying on the wood.

Now lay a flashlight cell between the lower ends of the two L-shaped supports. Let a pupil press these ends against the cell for just an instant. *What happens?* Then, by pressing and releasing the ends, he or she can start and stop the flow of electricity through the swing. *What does this make the swing do?*

What is the effect of turning the cell around, thus reversing the flow of electricity? What happens when the magnet is turned end for end?

BOBBY-PIN COMPASSES

Whenever possible, children should carry out science activities themselves, and not merely look on. For example, instead of just watching someone demonstrate a magnetic compass, pupils can make and use their own, as follows.

Straighten out a bobby pin, except for a small bend at the center. Then hold it by this bend, and stroke it from tip to tip with one end of a strong magnet. Do this at least 30 times. Be sure to stroke it along its entire length with the same end of the magnet—in one direction only. Keep the other end of the magnet away from it.

Does the bobby pin become magnetized? Can it pick up a paper clip or two? If so, tie a *single* piece of *thin* thread, or a long hair, to it at the bend. Then hang it up so that it can swing freely—at least several feet away from iron or steel objects, such as scissors, pipes, and metal desks. Get it to hang level by sliding the knot sideways, if necessary.

A whole class may magnetize bobby pins and hang them up, far apart. *Do they all point the same way when they come to rest?* To find out, everyone should stand up and extend both arms in the direction of his or her bobby pin. *After being spun around, do the bobby pins point the same way as before?* (If not, they should be magnetized more strongly by stroking them many more times with a strong magnet.)

One end of a freely swinging magnetized bobby pin will consistently point northward (shown by the direction of shadows made by the sun at midday). This is the *north-seeking* end. It may be marked with a tiny sticker.

A bobby-pin compass like this usually works especially well outdoors, far away from all iron and steel, if shielded from wind. It may be hung inside a large jar or jug, from a stick across the top. *Does this compass work as well if the jar or jug is filled with water? Does the bobby pin then swing as much as before?* This is why liquid-filled compasses are used on airplanes and ships.

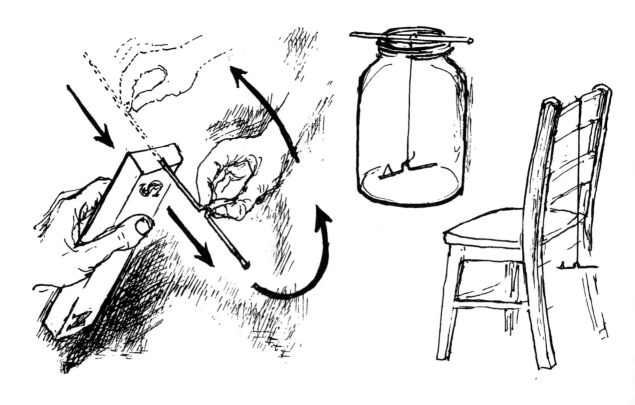

NEEDLE POLES

Pupils can learn a basic principle of magnetism at first hand by experimenting with magnetized sewing needles. They should *all* participate in this, as follows.

Hold a needle by its eye end and pull it across one end of a strong magnet—all the way from the eye to the point. Do this 30 times or more. On each return trip make a wide detour of the magnet.

Next, set an aluminum pan or glass bowl far away from any iron or steel objects. Pour in some water and float on it a piece of aluminum foil an inch or so across. Then lay the magnetized needle on the foil. *Which way does it point? When you turn it, does it go back to this position each time? If so, is this the way a compass points?* (See "Bobby-Pin Compasses," page 148.)

The end of the needle that points northward is a *north-seeking pole;* the other end is a *south-seeking pole.* The north-seeking pole should be marked in some way—perhaps with a dab of red nail polish.

Now take this needle off the foil. Magnetize a second needle, float it, and mark its north-seeking pole. Then bring the first needle close to it. *What do the two north-seeking poles do when near each other? The two south-seeking poles? One north-seeking pole and one south-seeking pole?*

These effects may also be tested with other pairs of objects that have been magnetized, such as:

Bobby pins (See "Bobby-Pin Compasses.")
Hacksaw or coping-saw blades (Suspend them by *single* threads so that they hang level.)
Steel slats from an old venetian blind (To make very large compasses, hang each one by a single thread tied to the middle of a yoke of string connecting the two holes.)

The earth is a huge magnet, with a North Magnetic Pole and a South Magnetic Pole. *Judging from the way the poles of magnetized needles affect each other, would you say that the earth's North Magnetic Pole is a* north-seeking *pole or a* south-seeking *pole? What does it do to the north-seeking pole of a compass? Then why is it named the* North Magnetic Pole? Let pupils use a globe to see where it, and also the South Magnetic Pole, are located.

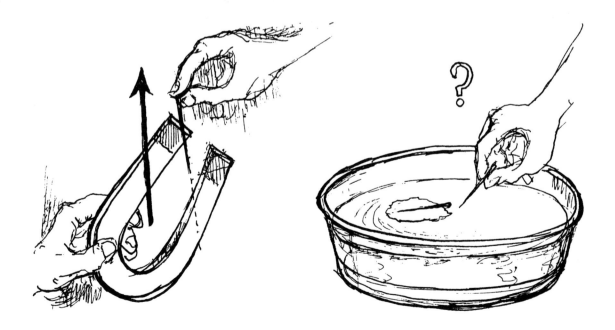

MAGNETISM RESEARCH

A good deal about magnetism can be discovered by pupils *on their own*. This provides great motivation—especially if they are allowed to research such questions as these:

1 *Is money magnetic?* Test coins, especially U.S. and Canadian nickels. Also test some bills. Hang one on a long hair inside a carton or jar, to shield it from air currents. Then slowly move a strong magnet close to it. Make no breeze!

2 *Are some sand grains attracted by a magnet?* Spread a little clean, dry sand on a sheet of paper and move a magnet beneath it. *If any grains are attracted, what might these be? Would they rust in water? When you test them, do they rust?*

3 *Can iron lose its magnetism?* Bend a tiny hook in one end of a 1-inch piece of very thin iron wire. Hang it on a pin in a block of wood. Set a magnet nearby, to pull the wire to one side without touching it. Then heat the wire red hot with a candle. *What happens? What happens when it cools?*

4 *How long will a homemade magnet keep its strength?* Magnetize a hacksaw or coping-saw blade as strongly as you can—as in "Bobby-Pin Compasses" (page 148). Now dip all parts of it into a pile of tiny nails and count all that cling to it. Find the average number and keep a record. Then retest the blade each week or each month.

5 *Will banging a magnet weaken it?* Test the strength of a magnetized hacksaw blade—as in #4 above. Then hit it hard—or drop it off a desk—10 times, and measure its strength again. Keep repeating this, make records, and draw a graph of the change in the magnet's strength.

6 *Do stationary steel objects have magnetic poles, and if so, where?* Test steel lockers and filing cabinets with a compass, top to bottom. Note which pole of the compass is repelled, and remember—*like* poles repel.

7 *Can you make a* one-*pole magnet?* Magnetize an old hacksaw or coping-saw blade—as in "Bobby-Pin Compasses." Then touch all parts of it to tiny nails; its poles are where most of these cling. Now try to magnetize a blade so that it has only *one* pole. *Can you do this? Do you get two one-pole magnets by breaking a magnetized blade in half?*

8 *Over what distance can a magnet act?* Hang a compass in a jug of water—as in "Bobby-Pin Compasses"—and lay a strong magnet near it. When it is still, watch it closely while someone turns the magnet end for end. Then try this with the magnet farther and farther from the compass. *How far must it be to have no visible effect?*

9 *Does anything block magnetism?* (See "Magnetic Tightrope," page 146.)

10 *Can copper or aluminum act as a magnet?* (See "Electromagnetic Swing," page 147.)

NAIL ELECTROMAGNETS

Electromagnets are extremely important in modern life—for example, for lifting iron and steel, producing sound in telephones and record players, and operating switches by remote control. However, their most important uses are for making electric motors run and for generating electric currents.

Even young children can have the fun of making electromagnets and experimenting with them. They need only wind insulated wire around a nail and touch the ends of the wire to a flashlight cell ("battery"). The winding may or may not be neat.

A flashlight cell, or even a few cells connected together, cannot give them a shock. *Caution: Be sure that nobody tries to connect an electromagnet to an electric outlet!*

All pupils should make their own electromagnets. For this, give them large nails, 3 inches or so long, and 8- to 10-foot lengths of insulated wire. The best wire for this purpose is #24 or #26 copper wire that is insulated with cotton thread or a coating of plastic. Such wire is sold as "magnet wire" by electrical supply stores.

Let each pupil wind a length of wire around a nail—the entire length of wire except for about 6 inches at each end. Help in scraping the insulation off the very ends of the wire, so that the metal is bare and shiny. Then show how to touch one end of the wire to the bottom and the other to the tip of a flashlight cell—*for a short time only*. During this time the nail will be able to pick up paper clips or other iron or steel objects.

An electromagnet should be connected to a cell for no more than a few seconds at a time. It uses the energy of the cell at a high rate, and will exhaust this energy if left connected for long.

(If suitable wire is not available, or if pupils want to try something different, suggest making an electromagnet with a coil of foil instead of wire, as follows: Cut a roll of aluminum foil to get a strip of foil 2 inches wide and 12 to 14 feet long. Also cut a roll of thin plastic wrap to get an equally long strip of plastic, but 3 inches wide. Now wrap a large nail in a single layer of foil that extends a few inches beyond the point of the nail. Then lay the foil strip lengthwise on top of the plastic strip, and wind both of them tightly around the nail. The plastic must insulate the turns of foil from one another, so that electricity will flow through the entire length of the foil strip, around and around the nail. Finally, tape the coil tightly, except for the two ends, and touch these to a flashlight cell.)

Now let pupils experiment with their electromagnets. *Are they as strong as other magnets? Can they act through plastic, aluminum, or rubber? Do they also have two different poles?* (See "Needle Poles," page 149.) *When disconnected, do they lose* all *their strength? If their connections to cells are reversed, do their poles reverse, too?*

At a later time, after everyone has made an electromagnet, pupils may see who can construct the *strongest* one. Give them a choice of a variety of

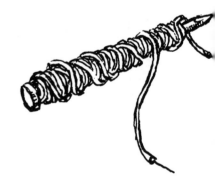

nails—even some of aluminum—and of wire. Then let each design and construct what he or she thinks will be the most powerful electromagnet that can be made with these things.

To test it, let its maker connect it to no more than three flashlight cells held end to end—*for only a few seconds at a time!* Then each pupil may list on the board the number of paper clips that his or her electromagnet can pick up at one time.

Another way to compare the strength of nail electromagnets is to lay an iron washer between two of them, so that the heads of both nails touch it. Then connect both electromagnets to the *same* cell at the *same* time, and pull them apart. *Which one holds on to the washer?*

Through firsthand experiences, pupils will be able to suggest factors that may affect the strength of a nail electromagnet, such as:

The thickness of the wire
The total length of the wire
The kind of insulation
The number of turns in the coil
The size or shape of the nail
The kind of metal in the nail—such as "soft" steel, "hard" steel, or aluminum
The number of flashlight cells, held end to end, to which the electromagnet is connected
The age of the flashlight cells, or how much they have been used

Then they can check each factor, one at a time, by a fair test in which all other factors are kept exactly the same. Thus, they might compare the strength of electromagnets made with identical nails and equal lengths of the same kind of wire, but with different numbers of turns in the coils around the nails.

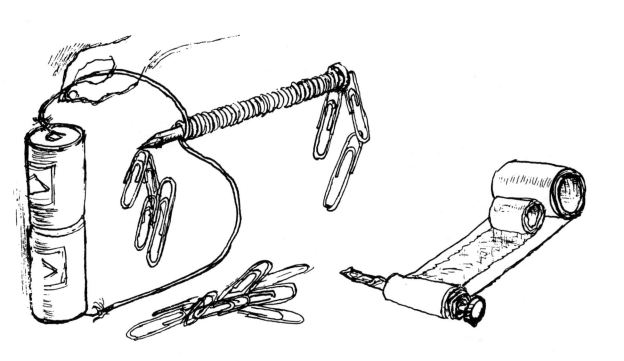

NIGHT LIGHTS

An excellent experience for children is to find out for themselves how to get a flashlight cell ("battery") to light a flashlight lamp ("bulb"). For this they should have little or no direction or help.

Pupils may bring the flashlight cells and lamps from home. They may also bring odds and ends of wire, and disposable aluminum pans and cans. Ideally, each child should have a cell and lamp, but they may take turns if necessary. However, each one should experiment *independently*. A demonstration would defeat the purpose!

Caution the pupils to stay away from electric outlets! However, they cannot get shocks from flashlight cells, unless many are connected together. A cell will not "wear out" quickly, unless its top and bottom are connected directly. And a lamp is not apt to "burn out" if it is connected to not more than one cell—or to two at most.

In how many different ways can you connect a lamp to a cell so that it lights? Does it light if you just touch it to the cell? Can it light without touching the cell at all? Can you use paper clips to connect it? Aluminum foil? What else?

Following this, pupils can make bedside night lights, in this way: Using adhesive tape or a rubber band, fasten a flashlight lamp firmly to one end of a long strip of aluminum cut from a disposable pan or can. Or fasten the lamp to a piece of fairly stiff, bare wire—with a screw-base lamp, simply wind the wire around the base, in the grooves. The metal strip or wire must make good contact with the base, but only on the side—not on the bottom.

Next, fasten the metal strip or wire firmly to the side of a cell with rubber bands or tape. Then bend it so that the bottom of the lamp presses against the tip of the cell. Also bend the metal strip or wire so that it touches the underside of the cell. Fasten it here with tape. Now the lamp should light—and a piece of index card slipped beneath it should make it go out. *Will a coin also do this?*

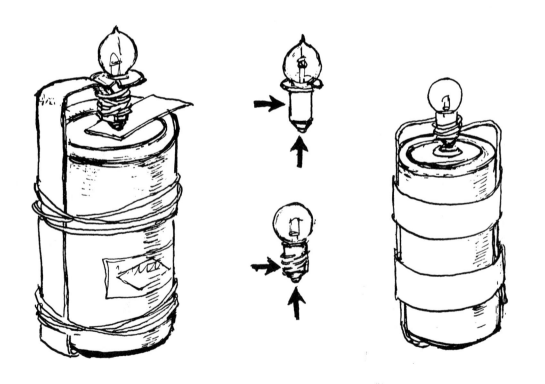

LIGHT CIRCUITS

An electric light such as a flashlight lamp (''bulb'') is lit by electricity flowing through it. For this to happen, it must be part of a continuous loop of materials through which electricity is made to flow. Such a loop is called a *circuit*.

A lamp is usually connected in a circuit by means of a *socket*. This holds the lamp firmly. At the same time, it connects pieces of metal in the base of the lamp to wires in the circuit. (See ''Night Lights,'' page 154.)

Sockets for flashlight lamps may be purchased from radio shops and electrical supply companies. However, pupils can construct their own in a variety of ways, as suggested by the sketches. They may use wire (preferably #18 or #20 copper ''bell'' wire) with the insulation scraped off, thin aluminum cut from disposable pans and cans, scrap pieces of soft wood, and thumbtacks.

Cell holders, for connecting one or more flashlight cells in circuits, may also be purchased.

Again, however, pupils can make their own, as suggested by the sketches below (and under ''Switches in Circuits,'' page 156).

With flashlight lamps and cells, lamp sockets, cell holders, and wire (#20, #22, or #24 copper ''magnet'' wire is ideal), pupils can connect simple light circuits. They cannot get shocks from flashlight cells—unless they connect quite a few together. ***Caution them to stay away from electric outlets!***

Then, by experimenting, they may investigate questions such as these:

1 *Must the plastic or other insulation be taken off wires where connections are made?*

2 *Will uninsulated wires work? Strips of aluminum foil?*

3 *Can electricity flow through coins? Through pencil lead?*

4 *Does it matter where a cell is located in a circuit?*

5 *How can two or three lamps, or two cells, be connected in a circuit?*

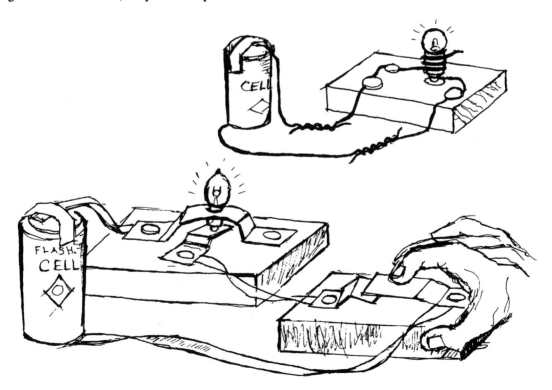

SWITCHES IN CIRCUITS

An electric light is usually turned on or off by a *switch*. This is simply a device that closes or opens a gap in an electric circuit. Closing the gap completes the circuit and allows electricity to flow; opening it "breaks" the circuit and stops the flow.

A simple switch consists of two pieces of metal that can be made either to touch or to stay apart. Such switches may be purchased from radio shops or electrical supply companies. Some are completely enclosed, and their parts are hidden; others have their parts readily visible.

However, as shown by the illustrations below (and under "Light Circuits," page 155), pupils can easily make their own switches. They may use fairly stiff wire or springy metal cut from disposable pans or cans, blocks of soft wood, and thumbtacks. Then they can connect these switches in circuits made with flashlight cells ("batteries") and lamps ("bulbs"). *Caution: They must not connect them to electric outlets!*

By experimenting, pupils can then answer questions such as these:

1 *Does it make any difference where a switch is connected in a light circuit? Will it work equally well on either side of the lamp?*

2 *How far can a switch be from a lamp and still control it? Can it be as far as 10 meters away? As far as 100 meters?*

3 *When is it better to have a switch that stays on by itself, and when is it better to have one that must be held closed?*

4 *How can a homemade switch be made so that one cannot possibly get a shock from it?*

5 *How can a lamp be connected in a circuit so that* either *of two switches can turn it on or off?* (This is illustrated by the lower sketch below.)

A light circuit with a gap in it can also be used to test whether materials are conductors or non-conductors of electricity. *For instance, can aluminum foil "bridge" the gap and allow electricity to flow? Can plastic? Pencil lead?*

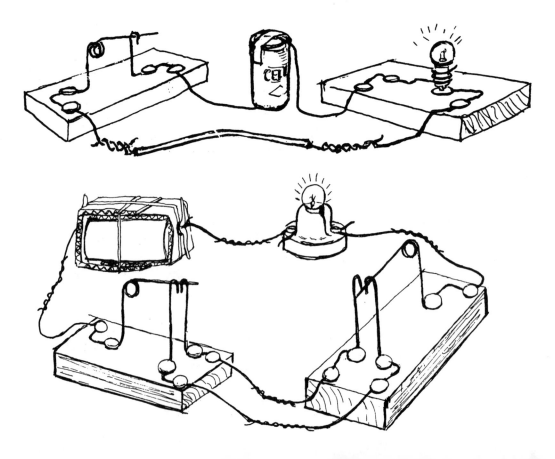

CHARGED CHILDREN

Many children know that rubber balloons, after being rubbed with wool, can attract hair, bits of paper, and other light objects. Such balloons are said to possess *charges of electricity*. (Our word *electricity*, interestingly, comes from the Greek name for amber, *elektron*. Ancient peoples knew that amber, after being rubbed with wool, could attract bits of straw. *What is* amber?)

Pupils can easily produce electric charges on many common materials. This works best when the humidity is low—when the air is "dry," as in a heated room during cold weather. The easiest way is to rub two different materials together. Besides rubber and wool, some that are likely to work well are plastic and fur, nylon and feathers, and a sheet of paper and a brush with natural bristles.

Children will be fascinated to find that they, too, can be given electric charges. To show this, ask a pupil who is wearing a wool sweater or leather jacket to stand on a short board resting on four cakes of paraffin. Then let someone rub the sweater or jacket briskly with a large plastic bag.

A child who receives a charge is able to attract various light objects hanging on threads—even a balanced yardstick—as well as a thin stream of water from a faucet. He or she can also cause a spark by touching someone's finger, or a flash of light by touching one end of a fluorescent lamp held by someone. The spark or flash of light shows up best, of course, in a dark place.

People often receive electric charges by sliding across plastic-covered automobile seats, scuffing their feet on rugs, or removing clothing made of certain fabrics. They may then cause sparks, receive mild shocks, attract light objects, or make fluorescent lamps flash.

Can persons wearing something other than wool or leather be given electric charges in this way? How about plastic jackets or raincoats, when rubbed with wool? Is the kind of charge always the same? (See "Kinds of Charges," page 158.)

KINDS OF CHARGES

As a sequel to "Charged Children," pupils may be guided through the reasoning behind the concept of two different kinds of electric charges—*positive* and *negative*. They can then test for these kinds of charges.

When the humidity is low, lay two 6-inch lengths of *woolen* yarn—*real wool*—on a desk. Rub them several times with a rubber balloon, from end to end. Then hold them up by one end, near each other. *What do they do? What happens when more than two pieces of yarn are used?*

The pieces of yarn are all of the same material. They have been given the same treatment. *Would you not expect them to have the same kind of electric charge?*

Now cut several long, narrow strips of rubber from a balloon. Rub them with a ball of woolen yarn and hold them up near each other. *What do they do? Are these strips of the same material? Have they been treated in the same way? And so, would you not expect them to be charged alike?*

Next, hold up a charged piece of woolen yarn near a charged strip of rubber. *What do they do? Do they act like objects that have the same kind of electric charge?*

Thus, when rubber and wool are rubbed together, they both become electrically charged—but with different charges. The charge on the wool is called *positive;* that on the rubber, *negative.* There seem to be no other kinds of charges.

Often, after pupils rub such things as feathers, rubber bands, or strips of newspaper with plastic, wool, or silk, samples of the same material repel each other. If so, they also repel either a charged piece of wool or a charged strip of rubber—not both. If they repel the charged wool, they have the same kind of charge as it has—positive. If they repel the charged rubber, they have a negative charge.

Knowing this, pupils can easily identify the kind of electric charge on a charged object. They need only note whether the object repels charged wool or charged rubber.

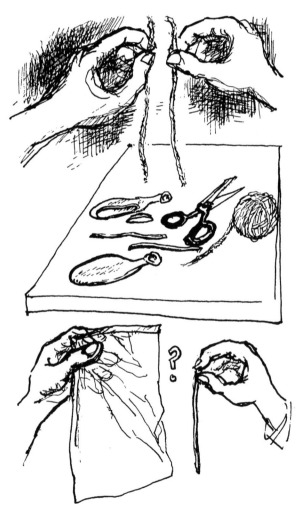

PIE-PAN GENERATOR

When the air is ''dry,'' as it generally is indoors in winter, children often make sparks and get slight shocks. This may happen, for instance, when they touch a doorknob after scuffing their feet on a rug. However, they can generate larger electric charges, and control them more easily, with a pie-pan generator, made and used as follows.

Stretch two rubber bands across a lightweight aluminum pie pan, at right angles to each other. Next, flatten out a plastic bag on a desk, rub it briskly with a woolen scarf or mitten, and hold it up by one edge. Then, without touching the metal, lift the pan by the rubber bands. Hold it against the bag, and while it is there, ask someone to touch it for just an instant. *What do you hear?* Now, without touching the metal, take the pan away from the bag and touch it to someone's nose. *What happens?*

The sparks show up better in a dark room or closet. Here pupils may also touch the charged pan to one end of a ''burned-out'' fluorescent lamp that someone holds by the other end. *What do you see?*

How a pie-pan generator works can be explained fairly simply. When the plastic bag is rubbed with wool, it becomes negatively charged. Then, when the pan is held near the bag, the negative charge on the plastic repels negative electricity that is in the metal. Some of this electricity jumps to a nearby finger, making a spark, and the pan is left with a positive charge. And so, when it is brought up to someone's nose, the pan attracts negative electricity from the person, causing a second spark.

Pupils can easily test the kinds of charges on the plastic bag and metal pan. (See ''Kinds of Charges,'' page 158.) Also, as a variation in the way the pan is charged, they may use a flat piece of woolen cloth or felt in place of a plastic bag, first rubbing it with plastic. *What kind of charge do you predict the pan will get then? Is this what an actual test shows?*

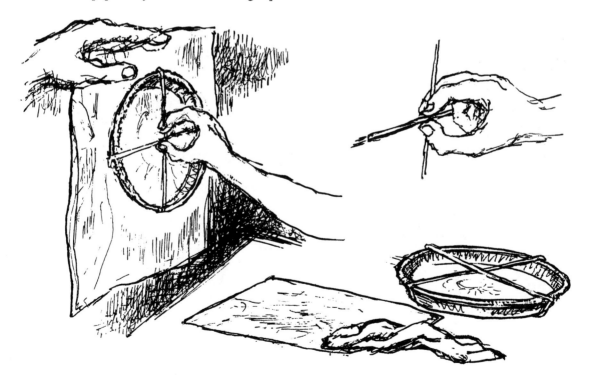

CHARGES AND CURRENTS

It is often stated that electricity is of two kinds, *static* and *current. Is this true? Are these really two different kinds of electricity? Or is this a little like saying that there are two kinds of water,* still *and* flowing? Pupils can investigate and find out.

When the humidity is low, as in a heated room during cold weather, rest a fluorescent lamp (a "burned-out" one usually will work) on a table in a dark place. Have one end touch a metal cabinet, locker, or pipe. Then produce an electric charge on a metal pan (see "Pie-Pan Generator," page 159) and touch the pan to the other end of the lamp. *What does the lamp do?*

Now scrape bare one end of a long wire, and fasten it to the prongs at the end of the lamp the pan touched. Run the wire through rubber bands, and let pupils standing in a line hold these—or hang them from chairs. The wire should touch nothing but the rubber bands and the metal prongs

at the end of the lamp. Then charge the metal pan again, and touch it to the free end of the wire, also scraped bare. *What happens? How can you explain this?*

As long as an electric charge stays on the pan, it is *static.* But if the charge moves, there is an electric *current. Do you cause a current in the wire? If so, can you tell its direction?*

Actually, an electric charge may flow in either direction through the wire, and still make the lamp flash. Which way it flows—the direction of the current—depends on the kind of charge on the pan. *What determines this?* (See "Pie-Pan Generator.") If the pan is charged positively, negative electricity flows *toward* it; if it is charged negatively, negative electricity flows *from* it.

How long can the wire be and still work? Can it be twice as long as the room? Do various kinds of wire work? How about string, rubber bands linked together, or a long strip of aluminum foil? Are rubber bands really needed to hold the wire?

DETAILS

Light and Other Radiations

Visible light is perhaps the most important form of radiant energy. Although it is only a tiny fraction of a great spectrum of radiations (or "waves"), which travel at astonishing speeds through space, light is the major means of our learning about things and of communicating with others. It enables us to read; to enjoy a landscape, a painting, a TV show, and the starlit sky; and to photograph these things. It also enables a green plant to make food.

Early experiences with light should include playing with shadows, mirrors, and magnifiers. Children enjoy projecting fantastic figures on screens, using mirrors to see around corners, and peering through lenses at all sorts of things. They are intrigued by magnifications, distortions, inversions, and other illusions they observe at air/water and air/glass surfaces.

In their study of light, children should become aware of that marvelous, yet delicate, instrument—their own eyes. They should find out how eyes are similar to, and different from, a camera,

and how to take proper care of their eyes. They should learn how to provide adequate lighting for reading and other close work.

Practical applications of these basic experiences should include holding objects to avoid glare and annoying shadows, and holding and using optical instruments such as a hand magnifier, a microscope, and a camera. Children should be able to infer the location of a light source and the shape of some objects from observing shadows. They should also be able to detect and explain some common optical illusions.

Besides visible light, the spectrum of radiant energy includes heat and radio waves. Simple studies of radiant heat energy require only hand magnifiers and thermometers. Children are familiar with televised close-up pictures of other planetary surfaces, radioed to the earth. The convenience and popularity of small, transistorized radios makes it possible for children to experiment with such radio waves, which are still the most widely used means of long-distance communication.

161

SOME IMPORTANT OBJECTIVES

Attitudes and Appreciations to Be Encouraged

Knowledge depends largely upon light, since direct observation is the basis for most of what we know about our universe.

Since we depend so much upon our ability to perceive light, our eyes should be given the best of care.

Just as some animals have a hearing range far beyond our own, there may be other animals or objects that are sensitive to radiations we cannot perceive.

Since most objects can be seen from many directions, space must be filled with light from a tremendous number of sources, traveling in all directions.

Two persons' observations of a single event differ because light, like any other sensation, must be perceived and interpreted, and no two persons will see the same thing in identical ways or from the same position at the same time.

Seeing is not necessarily believing, since light can be bounced and bent to produce illusions not obvious to an unwary observer.

Just as people differ in their sensitivity to sounds, they also differ in sensitivity to light, so that visual qualities such as color and intensity depend, in part, upon the person; they are not absolute.

What we see happening is always history, since light is not instantaneous, and the historical effect increases with distance.

Skills and Habits to Be Developed

Protecting one's eyes and those of others from injury and resisting the temptation to rub them if particles get into them

Determining the direction to light sources from observing the shadows cast by objects

Describing the form of an object from the shape of its shadow, and predicting the shape of the shadow an object will cast

Manipulating two variables (such as a pencil and a screen) simultaneously to produce a desired result, and then describing the degree to which each contributes to the result

Relating magnifying power to the degree of curvature of the surface of a lens or of a liquid-filled container

Making a magnifier from common materials such as water and wire

Holding and using properly instruments such as a magnifier

Holding a plane mirror to send a beam in a desired direction or to see objects not in direct view

Recognizing a common element in two situations that produce the same result—for example, that metal is common when both window screen and aluminum foil act as a barrier to radio waves

Using correctly such terms as *transparent, light source, plane, reflect, convex, focus, radiation,* and *field of view*

Facts and Principles to Be Taught

The closer an object is to a light source, the larger and less distinct is its shadow. The larger the light source, the less distinct the shadow.

Straight lines are the directions in which light beams tend to travel.

Things seen through a convex (bulging) surface of a transparent solid or liquid usually appear magnified, but things seen through a concave (depressed) surface appear smaller than they really are.

The more sharply curved the surface of a magnifying lens, the greater is its magnifying power.

When a beam of light strikes a reflecting surface, the angle at which it leaves the reflector equals the angle at which it strikes.

People looking at themselves in a vertical mirror can see twice as much of themselves as the mirror is long.

How much of other objects a person can see in a mirror depends upon where the objects, the person, and the mirror are in relation to each other.

Radio waves pass readily through glass, wood, plaster, and plastic, but they have difficulty passing through metal, even when it has holes in it.

SHADOW SHAPES AND SIZES

Young children love to pantomime animals with their fingers between a light source and a screen. As a teacher, you should capitalize on this interest to let pupils investigate what affects the size, shape, and sharpness of shadows.

With the help of the custodian, suspend a 200- or 300-watt light bulb near the center of the room. See that each pupil has a pencil, a sheet of white paper, and a cardboard on which to tape the paper to make a flat, stiff screen. Have available a supply of straight pins, some washers, and a few small wooden cubes or dice. Darken the room, turn on the lamp, and let the pupils proceed as follows.

Hold the pencil so that its shadow on the screen is the same size and shape as the pencil. Then hold the pencil so its shadow is a small, nearly round spot. (If finger shadows bother, stick a pin into the pencil to make an almost shadowless handle.)

Now try holding the pencil so its shadow is shorter, but no wider, than the pencil. *Are both ends of the pencil the same distance from the screen? Is one end of the shadow more distinct than the other? Which one? Can you hold the pencil so that its shadow is longer than the pencil, but no wider? What is the longest pencil shadow you can make? The widest?* Notice that inclining the screen, as well as the pencil, affects the size and shape of the shadow.

Hold the washer so that its shadow is nearly a circle. Hold it so that its shadow is a line about as long as the diameter of the washer. Try holding the screen and the washer to make a line shadow as long as the screen. *Is the shadow equally sharp at both ends?*

Can you hold the cube so that its shadow has more than six edges? Less than four? Can a rectangular solid have a shadow with five sides?

At home, hold a pencil under a single fluorescent lamp. *How does the shadow of the pencil held* parallel *to the lamp compare with one made when the pencil is swung at* right angles *to it? Can you tell why?*

BIG FINGER

Olives usually are sold in glass containers of small diameter. To find out why, each pupil can bring one of these empty bottles to school and follow the directions below.

Fill the bottle three-quarters full with water. Stick one index finger in the water, and hold the other outside the bottle to compare. *How does the one appear different from the other? Does the finger in the water appear larger when you hold it near the back of the bottle or near the front?*

Put a ruler in the bottle, or if the bottle is too narrow for a ruler, make marks 1 centimeter apart on a strip of paper and put it in the bottle. *Does the bottle make centimeters appear longer?* Cut two strips of notebook paper, one with lines running lengthwise and one with lines running across the strip. Put both in the bottle. *What happens to the spacing? What happens to the appearance of the squares on a strip of graph paper placed in the bottle?*

Now try these same activities with containers of larger diameter, such as peanut butter or mayonnaise jars. *In which is the appearance of objects changed most?*

Replace the jars with nonround containers such as an orange-juice bottle. *Does a pencil in a water-filled container of this sort look wider through the flattened side?*

Try the same objects in a clear shampoo tube whose top has been cut off to make the opening larger. Squeeze the tube and see what difference its shape makes in the appearance of things inside. *To make objects look as large as possible, in what kind of a bottle would you put them?*

Imagine an object in a water-filled olive bottle. Suppose that you wanted to know how wide it was, without removing it from the bottle. *Could you tell by holding a ruler in front of it? Why? Could you tell by immersing a strip of marked paper in the water next to the object? Why? Could you tell, if the object was spherical like a marble? How?*

LITTLE FINGER

In "Big Finger" (page 164) children learned how to make things look wider by putting them in a slender, round bottle filled with water. *What do you think would happen if the air outside the bottle and the water inside it were exchanged? If a water-filled bottle in air makes things look wider, will an air-filled bottle in water make things look narrower?* Let the children write their guesses, and then let them check their guesses by experimenting as follows.

Fill an aquarium with water to a depth about 2 centimeters less than the height of an olive bottle. Ask a partner to hold a clean, dry olive bottle upright against the bottom of the aquarium. Then stick a finger down inside the bottle. At the same time, hold the corresponding finger of the other hand in the water beside it. Ask your partner to observe and compare the two fingers. Now trade places with your partner. Observe and compare

your partner's fingers. *Does either finger appear magnified? Does either appear smaller? Which one? Where does the appearance of the finger inside the bottle change most—near the front of the bottle, or near the back?*

Try some other objects in the air-filled bottle. *Does the spacing of centimeter marks on a ruler or on a strip of paper seem to change inside the bottle?* Put a strip of squared (graph) paper in the bottle. *How does the apparent shape of graph-paper squares compare with what you saw in a water-filled bottle?*

Replace the slender olive bottle with a jar of larger diameter. Try your finger, a ruler, a strip of graph paper, and some other objects in it. *Can you see that things are not reduced in width so much as in the olive bottle?* Large-diameter jars filled with water do not magnify so much as small-diameter jars. And, in large-diameter jars filled with air things do not appear so narrow as in small-diameter jars.

WATER-DROP MAGNIFIERS

Pupils can feel that a magnifying glass is thicker at its center than at its edge, but they may not know what makes some magnifiers stronger than others. Given the following directions, however, they can find out for themselves.

Lay a sheet of newspaper on a table and put a sheet of waxed paper over it. With a medicine dropper, put a drop of water on the waxed paper. Slide the paper on the newsprint until the drop comes over a letter. *What does the drop do to the letter? Can you see the whole letter at once in the drop?*

A centimeter or two from the drop, make a larger water *lens* by putting five drops in a puddle. Slide the waxed paper until the puddle is over the letter you just looked at. *Does the puddle magnify the letter more, or less, than the single drop did?*

Make a puddle of 10 drops or more. Move this over the same letter. Compare what you see now with what you saw before. *Which of the three lenses is strongest? Which lens lets you see more of the letter, or more letters, at a time?* Try the three lenses on a newspaper picture. *In which lens do the dots of the picture look largest? In which lens can you see most dots? Do you think a lens can be a strong magnifier and let you see a large area at the same time?*

Bend down so that your eyes are almost at the level of the waxed paper and look at the three water lenses from the side. Compare the curvature (roundness) of the single drop with that of the two puddles. *Which has the sharpest curvature?* This is like the sharper curvature of a slender olive jar (see "Big Finger," page 164) compared with that of larger jars.

Bend the end of a thin wire around a large nail to make a loop. Catch a drop of water in this loop, hold it horizontally near a letter, and observe the letter. Carefully touch the drop with a finger and remove a little of the water from it. *What happens to the magnifying power of the water lens as it becomes thinner? Can you remove enough water to make it a* reducing *lens?*

DETAIL ENLARGED

SURPRISE IN A MIRROR

How tall does a "full length" mirror have to be? A surprise may be in store for the class and teacher who actually check what they think the answer is!

First hold up a rectangular mirror, ideally about 8 by 10 inches in size. *How much of yourself would you be able to see in this mirror? Could you see from the top of your head down to your chin? Down to your neck? Your waist? Your toes?*

To find out, pupils may work in groups, each with a mirror. One pupil should hold the mirror flat against a wall while the others, one at a time, stand where they can see themselves in it. Each should ask that the mirror be raised or lowered until the reflection of the top of his or her head appears even with the top of the mirror. *How far down on yourself can you see? Can you see more of yourself if you move farther from the mirror?*

Now the pupils should measure on themselves how far down each can see in the mirror. *How does this compare with how tall the mirror is? How tall would it have to be for you to see* all *of yourself?*

To check this, someone may draw a 6-inch tall figure of a person on a sheet of paper, and punch a small hole through one of its eyes. Then one pupil should view the figure by holding it in front of an upright mirror and looking through the "eye." Another should mark on the mirror, perhaps with tape, where the head and feet of the reflected figure appear to the viewer. *How much of the mirror is needed for the viewer to see the entire figure?*

To see *why* this is true, a pupil may draw a figure near one edge of a sheet of paper, and a line near the opposite edge to represent an upright mirror. Then the pupil should rule lines to show how light travels outward from the figure, in all directions. If light hits the mirror, it is reflected the way a ball would bounce off—at an angle equal to the angle at which it hits. *Where would light from the figure's foot have to hit the mirror to be reflected to the figure's eye? Where would light from the top of its head have to hit the mirror to reach the eye? How far apart are these two places? And so, how tall would the mirror have to be to be "full length"?*

SEEING DOUBLE

Most illusions are the result of light reflecting or refracting in a certain way. Children love to be fooled by an optical illusion, but when they can fool others, it is even more fun. An illusion that they can perform, while others watch, can be done with a large wall mirror, or with the store-front window in any store with a set-back front door.

Ideally, the wall mirror should be several feet high and wide. But since a mirror of this size is unlikely to be found except in such places as a hotel lobby, have groups of students try a store-front window instead. The performer should stand in the entrance to the store, facing along the street. The *midline* of the performer's body should be *in the plane of the front window*. The observers should stand several feet away, close to the window, facing the performer. Then the performer should proceed as follows.

While a friend holds your hand so you don't lose your balance (or while you grip the wall of the entrance), lift both an arm and a leg on the street side of the window. *What is the reaction of the observers? To them, what does it appear that you are doing? Can you do this with a window that has ads and other papers posted to the inside?*

Try holding a cup in the hand opposite the street side. Hold it so the midline of the cup is exactly in the plane of the window. Then lift the cup as if to drink from it. *To the observers, what does the cup appear to do?* Try this with a hat—and as you lift it, command the hat to do your bidding. Then raise it, and it will seem to obey, as if on command. *To the observers, what does the hat seem to be doing?*

Because the reflections in the window (or mirror) are brighter than any light that might come through it from the back side, the observers see only the direct and the reflected light from a raised arm and leg, or a raised cup or hat. The part of the hat that is behind the plane of the window is not seen. So whatever is reflected makes the lifted object appear doubled. The performer seems to lift *both* arms and *both* legs! And a cup or hat really seems to obey!

SPACE SIGNALS

Radio waves are the most important means of communication in space. Even though they are invisible, there are some things pupils can learn about them. Ask the class to bring in one or more pocket-size radios. Set one on a table where all can see and hear it. The radio station is some distance away, yet the waves it sends out are able to reach the radio. *Do they really pass through materials such as glass and plaster to do so? If other materials were used for windows and walls, would the radio still play?* Divide the class into as many groups as there are radios, and then let them experiment to find out, as follows.

Turn the radio on and set it inside a wide-mouthed gallon jar. Cover the jar with a pane of glass, or invert another jar over the opening of the first. *Does the radio still play?*

To find out if radio waves will pass through water, put the radio (still turned on) in a plastic bag that has been checked for leaks. *(How would you check it for leaks without getting it wet inside?) Can you still hear the radio?* If so, immerse the sealed radio in a gallon jar of water and listen carefully. *Do radio waves still reach the radio? How can you tell?*

Remove the radio and wrap it in a piece of aluminum foil. *What happens as the foil encloses the radio?* Open the foil to make sure the radio still is turned on, and then close it again. *Does the sound stop?* Cut a 1-inch hole in the foil. *Does the sound come out? What happens as you increase the number of holes?* Remove the foil completely. *What difference does it make? If you lower the radio into a metal pail, does it continue to play? What if you put it inside a plastic pail?*

By completely surrounding the radio with various materials, you can find out which ones block radio waves and which do not. Try window screen, rubber, cardboard cartons, and metal cans. *Does metal screen seem to interfere with radio waves, even though the screen has holes? Why do you think a car radio needs an antenna, but a house radio does not? Why does a car radio "cut out" on a steel bridge, even with an antenna?*

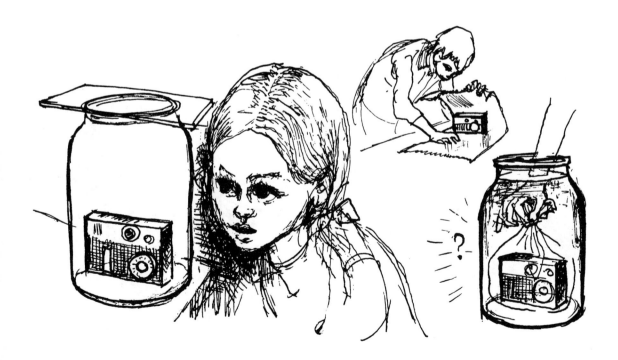

CARDBOARD-CARTON CAMERA

Long before moving pictures and television, people sometimes amused themselves by sitting in a dark room that had a small hole in its outer wall. Inside this *camera* (the Latin word for "room") the only light came from outdoors, through the hole. It produced a picture or *image* of the outdoor scene on the wall opposite the hole—a moving picture, in color, but upside down!

It is easy to make a camera of this sort for pupils to use and enjoy, or perhaps to copy in making their own. Start with a large cardboard carton that has been emptied but not torn. Paste a sheet of white paper inside it, against one end, and tape the carton completely shut. Then, in the bottom of the carton, near one corner opposite the white paper, cut a hole just big enough to admit a pupil's head. Finally, with a large nail, poke a hole through the end of the carton opposite the white paper. Locate it slightly above and to one side of where the pupil's head will be; light that comes through it must get to the white paper.

Now let each pupil, in turn, face away from a window that looks out on a sunlit outdoor scene, and lower the carton over his or her head. *Can you see a picture or image on the white paper inside? If so, is it right side up or upside down? Is it in color? Is it smaller or larger than the actual scene?*

Also, on a clear day when there are scattered clouds, let pupils go outdoors and take turns putting their heads inside the camera while *facing away from the sun.* **Caution them never to look directly at the sun!**

Can you see the sun's image on the paper? If so, how big is this image? What happens to it when a cloud passes in front of the sun? Does the image of the cloud move in the same direction as the cloud is moving?

Cardboard-carton cameras like this provide an excellent and safe way to view an eclipse of the sun. For this purpose each viewer should make his or her own, well ahead of time!

Pupils may also investigate the images made by an adjustable "pinhole camera," as follows:

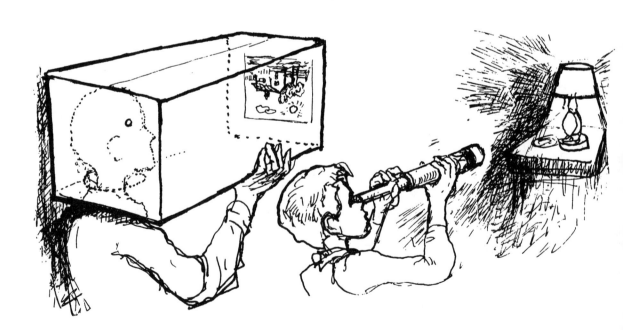

1 Find two cardboard tubes about 25 to 30 centimeters long, one of which is just able to slip inside the other. Try mailing tubes or tubes from inside rolls of paper towels, waxed paper, or aluminum foil.

2 Fasten a flat piece of waxed paper, tracing paper, onionskin paper, or translucent ("frosted") plastic across one end of the narrower tube. Tape it on, or use glue or cement spread on the edge of the tube, and then trim off any excess paper or plastic.

3 Similarly, fasten a flat piece of aluminum foil across one end of the wider tube. Then make a pinhole in the center of the foil.

4 Slip the narrower tube inside the wider one, with the open ends of both toward you. Then look into the inner tube, close your other eye, and point the tubes toward a bright window, an electric light, or a lighted candle.

Can you see an image on the translucent paper or plastic? If so, is it right side up or upside down? In color? Smaller or larger than the actual scene? Does the image of someone walking by move in the same direction as the person is walking?

5 Now, while looking into the inner tube, slowly slide the outer one back and forth. *How do the size and brightness of the image change?*

6 Later, make a second pinhole in the aluminum foil, about 1 centimeter from the first. *How does this affect the image? What happens when you rotate the outer tube? The inner one?*

7 After this, enlarge one of the pinholes with a needle. *What effect does this have on the brightness and sharpness of the image?*

At another time, pupils might replace the aluminum foil with a piece of clear sheet plastic, or fasten the plastic to the end of another, similar tube. Then they should make a tiny dot, preferably with black India ink, at the center of the plastic, and slip the tubes together as before. *When you look through the tubes toward an electric light, can you see the shadow cast by the black dot? If so, does it look the same as the shadow of the dot when the tubes are pointed at a bright window? At a lighted candle? What do you see when the light comes from two or more electric lights or candles? From a fluorescent lamp?*

Since the dot *stops* light instead of letting it pass through, it acts as a "negative" pinhole. Like other opaque objects, it casts shadows, and these are different under different conditions (see "Shadow Shapes and Sizes," page 163).

Heat and Energy

Heat and temperature are mysterious matters for many children. Hot and cold have sensible meaning for them, but to "have a temperature" usually means to be ill. And to many, adding a glassful of water at 10°C to another at 20°C suggests a mixture at 30°C! Children do not understand how a pitcher of ice water can have more heat than a steaming cup of coffee, or why opening an outside door in winter lets the heat out, not the cold in. It is important, in a world so dependent upon heat energy for transportation, industry, home comfort, and cooking, that children understand heat and how to use and control it.

Basic experiences should include investigations of hot and cold objects and practice in measuring their temperatures; of how materials vary in their ability to conduct heat; and of some common sources of heat such as friction, electric currents, burning, and chemical action where there is no fire. They should learn to respect, but not to fear,

heat sources and to use them safely.

Pupils should learn that the amount of heat in an object rarely remains the same for long but is ever being added to or lost. The rate of addition or loss of heat can be controlled, however, and they should learn ways of doing it and be able to select and arrange materials for the purpose.

They should learn the difference between heat and temperature, that one is related to the kind and amount of stuff in an object and the other is merely an indication of how hot it is. They should learn, too, that cold is the absence of heat and that an object feels cold only as heat moves into it from one's body.

By learning about heat and temperature, children will apply understanding to keeping food hot or cold, to dressing comfortably, and to holding hot or cold objects without discomfort, as well as to appreciating the extremes of the environment to which other living things are subjected.

SOME IMPORTANT OBJECTIVES

Attitudes and Appreciations to Be Encouraged

Without energy from the sun to produce heat on earth, life as we know it would be impossible.

Humans are not at the direct mercy of the weather, because they have learned to insulate their bodies and their food and to control the temperature of their buildings.

Many other organisms must endure extremes of temperature not felt by humans because of their clothing and temperature-controlled homes, factories, and schools.

Conveniences such as refrigerators and stoves are such an accepted part of our life that we tend to forget our dependence upon them until a power failure makes it clear.

Some people in our country and millions in other countries still have only relatively primitive means of heating homes and cooking, and so endure hardships unknown to the rest of us.

Touch is not a reliable way of sensing the temperature of objects; it confuses temperature and conductivity.

Heat is an unavoidable and usually wasted by-product of energy transformations such as from electrical energy to light; as the cost and need for heat energy rises, ways must be sought to minimize or make use of this waste.

As human beings make increasing demands for heat energy on a decreasing supply of fossil fuels, alternative energy sources will have to be developed, along with a social acceptance of less convenience and comfort.

Skills and Habits to Be Developed

Reading a thermometer to a single scale division

Holding matches properly to strike them and prolong their burning

Selecting suitable nonconductors with which to hold and support hot or cold objects, or to keep food hot or cold

Being careful not to leave in direct sunlight objects that might be harmed by the absorption of radiant energy

Recognizing friction as a source of heat, and using it when needed or minimizing it when heat is unwanted

Opening a window slightly when a car is parked in sunlight

Minimizing waste of hot water in baths and showers, because of the great amount of energy needed to make the water hot

Allowing room in containers for liquids to expand as they warm, such as gasoline in a tank

Using containers of water to buffer temperature changes, such as in a homemade incubator

Using correctly terms such as *temperature, heat, radiant energy, absorb, conduct, insulate, expand,* and *mass*

Facts and Principles to Be Taught

Most objects tend to expand when warmed, and contract when cooled, the extent depending on their dimensions, material, and change of temperature.

The rate at which radiant energy from the sun is absorbed by an object depends upon its color, composition, and angle to the sun.

When water absorbs or loses heat, its temperature changes less than that of almost any other substance.

Metals are good conductors of heat, but most other substances—such as glass, wood, plastic, fur, and fabric—are not.

Dark-colored objects tend to absorb and radiate heat energy better than do light-colored ones.

Heat energy always moves from warmer to cooler objects, never from cooler to warmer ones.

Whenever heat is lost by one object, it is gained by another.

Long objects made of metal (bridges and steam pipes, for example) have provisions for allowing them to expand and contract with temperature changes; sometimes they creak as they do.

A large mass of a substance has more heat than a smaller mass at the same temperature; it may still have more heat even if the smaller mass is warmer.

HEAT TRAPS

Along with light, the sun gives off radiation that we cannot see. Some of this raises the temperature of surfaces that it strikes; hence, it is commonly called *heat radiation* even though it has no temperature until it is absorbed. To show how various surfaces differ in their ability to absorb this radiation, take your pupils to a nearby parking lot on a warm clear day to investigate the temperature of cars that stand in sunlight.

Ask pupils to feel the top of the hood of a light-colored car. Then let them try the hoods of black or dark-colored cars. *How much of a difference is there—a slight difference, or a great one?* Now have them test cars of a variety of colors. *How do their temperatures compare, according to the touch test? Does the position of the surface with respect to the sun's rays make a difference?*

Get permission to place thermometers on the seats of both light- and dark-colored cars. Leave them for a half hour. Be sure that the windows are closed and that the thermometers are not in direct sunlight, or they will not indicate the air temperature. Have the pupils read the thermometers as soon as the doors are opened. *In which cars is the air hottest? How hot?*

On another warm, sunny day take your class to a place where a blacktop drive meets a concrete walk. Let pupils who wish to do so remove their shoes and stand on the blacktop, then on the concrete. *Is the concrete warmer or cooler than the blacktop? How does this difference compare with the differences in temperature you observed inside cars in the parking lot?*

Since dark-colored objects ordinarily are better radiant-heat absorbers than light-colored ones, snow usually melts more quickly on blacktop roads than on concrete ones. *If you lived in the tropics where there was no snow, would you prefer a light-colored or a dark-colored roof on your house? Why?*

ICE-MELTING CONTEST

Children often use special containers that keep things hot or cold by retarding the flow of heat to or from them. To understand some ways in which the insulation is accomplished, they can have an ice-melting contest that is both fun and instructive.

First, let the pupils pair off and go outdoors. Give each team an ice cube of equal size. At a signal, let them play ''Melt-the-Cube,'' each team bringing heat to its cube by any means except smashing the cube or applying artificial heat. One pupil, or you, should act as timer. The first team to melt its cube completely wins the game. Afterward, return to the classroom to discuss the results.

How many used the heat of their bodies to melt their ice? How many rubbed their hands to warm them before holding the ice cube, or rubbed the ice directly on their clothing? Did any think to use dark (and hence often warm) *objects?* (See ''Heat Traps,'' page 174). *What procedure proved most effective?* When the pupils have exchanged observations, let them run the contest again to see how much improvement each team can make over its first try.

At another time, hold a ''Keep-a-Cube'' contest. Then the class can investigate ways to retard the melting of ice by the use of materials collected beforehand, such as paper towels, aluminum foil, cloth, and plastic. Let each team take materials to a desk and think about how it will proceed. Then as soon as you give out the ice cubes, shout ''Go!'' and note the time. Each team should wrap up its ice cube. From time to time during the day, a team may unwrap its package. If ice still remains, they should note the time and rewrap. If no ice remains, their official finish time is the last time they *did* see ice. The last team to finish wins.

Can any team keep its cube for an entire school day, without taking it out of the classroom? What material, or combination of materials, proves to be the best insulator? See if the pupils can tell why.

HEAT TRAVEL

It is often advantageous when heat travels readily through a material, such as the metal of a cooking pot. In many instances, however, it is better if heat does *not* travel readily, as through the handle of the pot. The difference is one of *conduction*.

Pupils can easily test various common materials to find out which ones are good *conductors* of heat and which are not. For this, divide the class into small groups. In each group appoint a "Candle-keeper" to set a candle securely in a metal pan, light it safely, and be responsible for it. ***Caution: Have a can of water near each candle, for use in case of emergency.***

Now give each group a 3-inch length of thick iron wire, perhaps cut from a coat hanger, and an equal length of copper wire of the same diameter. This may be scrap electric wire with the insulation removed. *Which wire is the better conductor of heat?* A pupil can test both wires simultaneously by holding one end of each and heating its other

end in the candle flame. Ask them to do this *independently*—not to share their observations until after everyone has completed the test.

In the same way, pupils can test matched pairs of wires or narrow strips of various other metals, such as aluminum and iron, or aluminum and copper. Scrap wires and strips of metal may be obtained from electricians, sheet metal shops, and junkyards. A custodian or a parent can easily cut pieces enough for the class.

Pupils can also compare how well heat travels through wood and iron by testing pieces of swab sticks and nails of the same size. Similarly, they can test the heat *conductivity* of pieces of glass rod or tubing obtained from a high school science department. *How do these three materials—wood, iron, and glass—compare in heat conductivity?*

Finally, let pupils see how well they agree on the order of heat conductivity of all the substances they tested. *Which is the best conductor of heat? Which is the poorest? Of what practical value are these facts?*

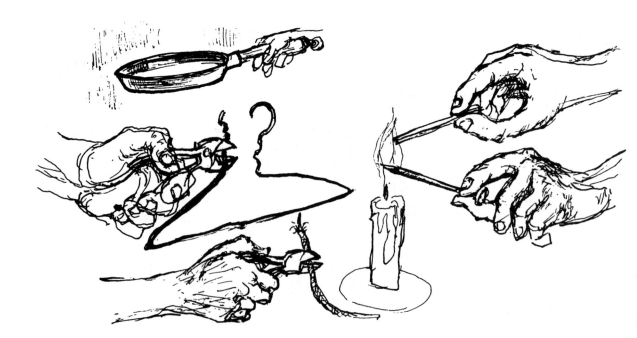

CURTAIN-ROD BRIDGE

Metal bridges expand or contract with temperature changes, and to allow for this they are usually set on rollers. What takes place can be shown with a model bridge made from an old metal curtain rod.

Support it, or some other straight metal rod a few feet long, on a pair of smooth wooden blocks. Tape these to a table to keep them from sliding, and tape one end of the rod firmly to one of the blocks. Then stick a straight pin through a narrow pointer cut from an index card or a drinking straw, and insert the pin under the untaped end of the rod. The pin must be able to roll easily, and the pointer to turn with it.

Now set several short candles under the "bridge" and light them. *What does the pointer do? Why? What does it do when the flames are blown out? When snow or crushed ice is sprinkled on the "bridge"?*

Bridges are not normally heated by fires, of course, but they are affected by air temperature and by sunshine. To show this, glue the two "piers" (wooden blocks) to a length of two-by-four, and set up the "bridge" without candles. Then carefully place the whole setup where sunshine can fall on it. Or set it outdoors on a cold day, and later bring it in.

How much does the "bridge" expand or contract? One full turn of the pointer would mean a change in length by an amount equal to *twice* the distance around the pin. (To make this clear, lay a stick across a cardboard tube on a table. Push the stick ahead until the tube makes one full turn. Then compare how far the stick moved forward with the distance around the tube.)

How can you find the distance around the pin? Suggest rolling a metric ruler across it as it lies on a block, with the pointer over the edge. *How far must the ruler move,* with respect to the pin, *to have the pin make 10 turns? And so, how far must it move to have the pin make* one *turn?* This is the same as the distance around the pin.

On the basis of what the "bridge" does, can you say, for sure, that all metals expand when heated? If not, why not? What can you say?

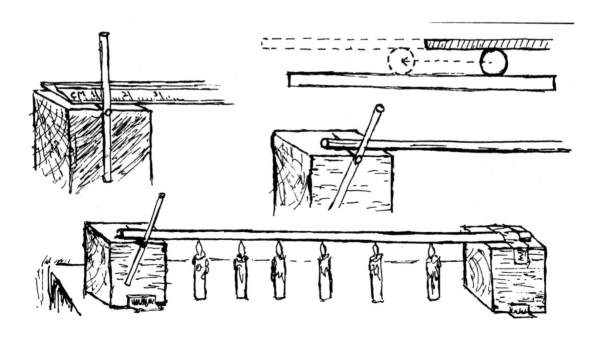

SHRINKING AND SWELLING AIR

In ''Curtain-Rod Bridge'' (page 177), pupils can see how much certain metals expand and contract with changes in temperature. By observing what happens to air samples as the temperature changes, they can compare the changes in gases with the changes observed in metals.

Divide the class into small groups, each with a round balloon, a quart jar, an inexpensive thermometer that fits inside the jar, some rubber bands, and several feet of string. Then let each group investigate the changes in air, as follows.

Inflate the balloon and seal it with a rubber band. *What is the approximate temperature of the air inside?* Now find the distance around the middle of the balloon and record it, either by marking the string or by measuring its length. Also, mark with a felt pen where the string was held against the balloon. Then set the balloon where the air temperature is much warmer or much colder than the air used to inflate the balloon. Leave it for a half hour. Record its temperature. (By that time the air in the balloon should be near the temperature of the air around it.) Holding the string on the balloon *at the same place as before,* measure and record the distance around the balloon. *How has the balloon changed in size? Approximately how much did it change in temperature? Is this change greater or less than for metals?*

Put the thermometer inside the jar. Cut the half containing the neck from the balloon, pull the rest of the rubber down tightly over the top of the jar, and fasten it in place with rubber bands. *What is the temperature inside the jar?* Cool the jar by putting it outside or in a refrigerator for a half hour. *What happens to the rubber as the temperature inside the jar changes? Where could the jar be placed to make the rubber cover bulge up?* Try it! When you do, however, remember to record both the temperature inside the jar and the shape of the cover.

PAPER POT

Most children know that paper and cardboard will burn when held in a flame. They may not know that each material must be raised to a particular temperature before it will ignite. For paper, this temperature is considerably higher than the temperature of boiling water. It amazes them to observe that water can be heated over an open flame in a paper "pot" without the paper itself burning.

Provide each group of about five pupils with the following materials:

Two small, "empty" juice cans
A square of duplicating paper, 12 centimeters on a side
A square of ½-inch hardware cloth, 15 centimeters on a side
A stout candle about 3 centimeters shorter than the cans
A sheet of aluminum foil, or a cake pan, to protect the table or desk top
A paper stapler or four paper clips
Thermometer and match

Then let each group proceed as follows.

Light the candle, drip some wax in the center of the foil or pan, and stand the candle upright in the drippings. Then make a rectangular pot by folding the paper square 4 centimeters from each edge. Fold the square at each corner diagonally, wrap the sides around the ends, and staple them in place or hold them with paper clips. Add water to a depth of 1 centimeter. Use the two cans to support the square of hardware cloth (screen) over the candle flame. Try to place it so the flame comes up in the center of one of the screen openings. Then set the paper pot, with its water, on the screen.

Have the pupils take the temperature of the water immediately and at five-minute intervals. They can also test the temperature with their fingertips from time to time. *Does the pot show any signs of burning?* (Sooty deposits from the candle do not mean charring of the paper.) *Whose pot is the first to show bubbles forming at the bottom? Whose is the first to "steam"?*

So long as there is sufficient water in the pot, the candle flame cannot heat the paper to its *kindling* temperature. The water limits the temperature to which the flame can raise the paper, even though heat is added to it and the water.

Caution: This should not be tried on a kitchen stove or a hot plate. Too much heat can be dangerous.

HEAT MIXTURES

The difference between heat and temperature is not clearly understood by most people. They understand the meaning of *hot,* and they know what it means to *heat* something. It is not clear to them, however, how one container of *warm* water can have more heat than another container of *hot* water. Some simple activities with water mixtures will help to show a class how this can be.

Divide the class into groups of about four children. Give each group a laboratory thermometer or a thermometer with a thin (and hence not massive) scale, three 250-milliliter and one 500-milliliter Styrofoam hot drink cups, and a clean, large milk carton whose top has been cut off. At opposite sides of the room set three plastic (*why not metal?*) pails of water—one cold, one warm, and one hot but not scalding. Then let each group proceed as follows, with each pupil making and recording all observations.

Fill one small cup about three-quarters full with cold water. Make a pencil mark on the cup at the water level, and pour the water into a second small cup. Mark where the water level comes, and pour the water into a third cup. Again, mark the water level. The marks on all three cups indicate equal measures. One cup still holds the cold water.

Now refill one of the emptied cups to its mark with *warm* water. Carefully take the temperature of the water in both cups. Then, without showing it to other groups, write a prediction of what the temperature will be when both cups of water are mixed in the large cup. Next try it and check your prediction. *How does the temperature of the mixture compare with the temperature in the cold and warm cups? How close was your prediction to the actual result?*

Pour out the mixture and try it once more. This time use samples of cold and *hot* water. Record their temperatures, write your prediction about the temperature of the mixture, then test to find out. *How does the temperature of the mixture compare with that of the samples? How close was your prediction? Can you see that when equal masses of*

water are mixed, the temperature of the mixture is midway between the temperature of the samples? One is just as effective as the other in determining the result.

Now put two measures of *hot* water in the large Styrofoam cup. Put one measure of *cold* water in a small cup. Take their temperatures, record them, and write your prediction of the mixture. Then pour them into the "empty" milk carton and find out for yourself. *How does the temperature of the mixture compare with what you observed in the cups? How close was your prediction?*

Repeat the activity, but this time use two measures of *cold* water mixed with one of *hot* water. *Can you see that the temperature of the mixture depends not only upon the temperature of the samples, but also upon their masses? A mass of water*

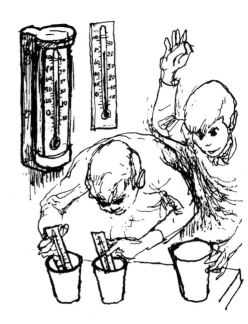

twice as great as another is twice as effective in determining the final temperature.

To test pupils' understanding of this principle, and to demonstrate that a large mass of water can have more heat than a smaller but warmer mass, try this. Fill one of the small cups to its mark with water at 0°C. (To get water at this temperature, half fill the milk container with cold water, add plenty of ice, and stir until no more ice will melt.) Also, fill a second cup to the mark with water at 50°C. Let two pupils check the temperature of the water in each cup and, when they agree on the temperatures, write them on the board. Next, ask each one in the class to write down his or her prediction of what the mixture will be when the two cups of water are mixed. Pour the two into the large cup and let the same two pupils check

the temperature. *How many of the class predicted it to within 2 degrees?*

Now refill a small cup to its mark with water at 0°C. This time fill one of the pails with, say, eight of those same measures of water at only 40°C. The water in the pail will be cooler than, but have eight times the mass of, the hot water just used.

Again, have these temperatures checked and recorded. Then ask the class what will happen when the larger but cooler (by 10 degrees) mass of water is mixed with the ice-cold one. Let the same pupils check the temperature and announce it. *How many predicted it to within 2 degrees? Which mass of water—the smaller, hot one, or the larger, but cooler, one—produced the warmer mixture?* Let the pupils who came closest in their predictions explain how they figured this out.

HEAT RESERVOIR

Suppose that equal masses of water and iron are at the same temperature. *If equal amounts of heat are added to each mass, would the temperature of one change more than that of the other? Which one?* Suppose, instead, that equal amounts of heat are removed from each mass. *Will one's temperature change more than the other's? Which one?* These are questions whose answers children do not usually know, and are surprised to learn.

Set out two pails of water, one hot (about 60°C) and one cold (near 0°C). Divide the class into groups of four or five each and provide each group with the following materials:

3 large insulated cups (one for carrying water)
20 8-penny nails
1 meter of slender string
1 equal-arm balance
1 thermometer of little mass (*Why little mass?*)

Then have each group proceed as follows.

Place one large cup on each pan of the balance. Tie the nails in a bundle and lower them into one cup. Add cold water to the other cup until both cups are balanced. This means there are equal masses in each cup. *Would that be true if the cups were not identical?*

Set the cup of cold water on the desk. Beside it set the cup of nails. Add hot water to the cup of nails until the water is at least 2 centimeters deep over the nails. After a minute, take the temperature of the water in each cup and record it. *Is the temperature of the nails the same as the temperature of the water around them? If not, what must happen to the nails? To the water? If the nails are lifted from the hot water and lowered immediately into the cold water, what do you think will be the temperature of the mixture?* Write a prediction.

Now lift the nails from the hot water and lower them into the cold water. When the temperature of the mixture stops changing, record it. *How many degrees did the temperature of the nails*

change? The temperature of the water? How close was your prediction?

Dry the bundle of nails with paper toweling, then repeat the activity. This time, balance the cup of nails with a cup of *hot* water instead of cold. And lower the nails into *cold* water before transferring them to the hot water. As before, take the temperatures in both cups before making a prediction. When the temperature of the mixture stops changing, take and record it. *Which changed more in temperature, the nails or the water?*

Because nails are made of iron, and iron seems heavy and hard to children, they think that somehow it should exert more influence than water. Water, however, has an amazing ability to absorb heat or to give it up, with little change in temperature. That is why water is often used to cool things fashioned from red-hot iron, and why water is the most common coolant in car engines. A sizzling hot frying pan plunged into dishwater causes little change in the temperature of the water, but the hot pan cools a great deal.

Similarly, water can give off a great deal of heat without its temperature changing very much. Suppose that you were cold when you went to bed, and you wanted something next to you that would stay warm for a long time. *Would you prefer a rubber container filled with hot water, or one filled with an equal mass of iron bits that were as hot as the water? Why?*

The temperature of mid-ocean islands tends to remain almost constant throughout the year. *Why should this be so? Why are animals living in water subject to smaller temperature changes than those living on land?*

It is fun and instructive for a class to set up a chick incubator and observe the changes that occur within fertilized eggs. As a precaution against cooling of the eggs if one of the light-bulb heat sources should "blow," jars of warm water may be kept beside the eggs. *Why is water better than iron for this purpose? Would a warm stone be as good as water?* Using the technique described above, the class has a way to find out.

BURNING UP CALORIES

Children hear about food having calories, and they may also hear about burning up calories while working or playing. They may have some notion that calories are related to energy, and that some foods that provide energy also have a lot of calories. How calories are able to help a person work or play, and how calories come out of food that is eaten, is difficult or impossible to show. However, you can show children that the energy in some foods will actually support burning, even though those foods aren't burned in quite that way in our bodies.

Divide the class into small groups, each with a foil pan, a lump of modeling clay, a toothpick, a few English walnuts, and some safety matches. Then let each group proceed as follows.

Stick a lump of clay to the inside of the pan, near the center. Carefully open a walnut and remove as large a piece of the nutmeat as you can. Stick the pointed end of a toothpick into this piece. Then stand the toothpick upright in the lump of clay, and pour water into the pan to a depth of about 1 centimeter.

Next, light a match and hold the flame under the nutmeat for a few seconds. *What happens to the nutmeat? Does the outside of the nutmeat seem to burn like paper, or does it seem to burn because of something in the nutmeat? Is the flame more like a candle flame, or like the flame from a gas burner? Is it a clean flame, or a sooty one? How long before the flame goes out?*

It is the oil in the nutmeat that burns. Many kinds of foods are rich in oil—not oil like engine oil, but oil such as vegetable oil, safflower oil, and corn oil. These oils will burn, as demonstrated with the nutmeat from the English walnut. So will many fats such as those in meat. The energy that they give off when burning is apparent, but the energy they give off when eaten is not so apparent. It is this energy, however, produced without a flame, that supports the work and play of pupils.

Let each group eat the rest of its walnuts, and some time later engage in some active exercise such as running up steps or running around the playground. Discuss with them how part of the energy that they needed for running came from the walnuts that they ate—energy that comes from flameless use of the calories by their bodies.

WASTED WATTS

People often think of electrical energy as "clean" and "efficient." Perhaps it is, compared to some other forms of energy, but our use of it may not be so clean and efficient as we sometimes think. It is wasted at every turn, and the wasted watts become heat, which is one kind of environmental pollutant. Some simple observations, and subsequent discussion, may help children to understand this, and to think about ways of using all forms of energy more efficiently.

For a class demonstration, get a hot plate, a teakettle or open pan, and a thermometer. Fill the kettle with water to a depth of about 3 centimeters. Then, before turning on the hot plate, take the temperature of the water, the air near the kettle, the surface of the hot plate, and the cord near the hot plate. Next, turn on the hot plate and put on the kettle. After the water is near boiling, turn off the hot plate. *Where has the heat energy (from the electrically heated coil) gone besides into the water?*

Which of the following things has been warmed, and about how much?

Surface of hot plate	Kettle metal
Base of hot plate	Air around hot plate
Water in kettle	Cord to hot plate

A careful examination of these things may show that all have been warmed, some much more than others. The base of the hot plate may have been warmed so much that it cannot be handled right away.

Only the energy that warmed the water was useful; the energy that warmed other things was wasted. It had to be paid for, though, no matter what it warmed or how it was wasted. *What suggestions can you provide to minimize this waste?*

A lot of energy gets wasted in a hot shower or bath, too. *Is it efficient to warm the walls of the hot-water heater? The air around the heater? The pipes that carry hot water to the bathroom? Is it efficient to heat the walls of the shower? The tub? The air in the bathroom? The water that runs when it is not being used? What are some ways to minimize wasted watts when taking a hot shower?*

The gasoline in a car does a lot of things besides move the car. *What are some of these things? Does a driver pay for moving air out of the way, as well as moving the car? How about a truck? What might be one important reason for using more fuel as speed increases?*

We need energy to live, but we must find ways to use it more efficiently. One way is by not demanding so many little extras from it. They all must be paid for in the end!

Sun, Moon, and Stars

Since prehistoric times, people have been intrigued by the bright objects in the sky. This wonderment still continues today, even with our highly sophisticated understanding of what these objects are and why they look, and change, as they do. How fortunate it is that human curiosity is not yet satisfied; that, with telescopes, space probes, and other instruments, the beautiful search for answers goes on—more energetically than ever!

Children are fascinated by the sun, moon, and stars. Their interest, moreover, becomes even keener when they have the opportunity and encouragement to observe these objects at first hand. Why must their learning be confined strictly to the classroom and to traditional time slots? Why should the emphasis be placed on mere memorization, instead of actual observation?

Fundamental to a person's comprehension of the universe are direct sensory experiences. These each child has a right to have for himself or herself. Pupils deprived of such experiences—though they memorize the distance to the moon, the size of the sun, and the names of the planets—get little real appreciation of astronomy. Nor can they acquire this appreciation merely from words, pictures, and models.

A telescope is not essential for teaching basic astronomy. Neither is a planetarium, a transparent globe, or a solar-system model. The teacher does not have to know a great deal of subject matter. What *is* needed, above all, is a willingness on the teacher's part to go outdoors—first alone, and then with the class—to observe the sun, moon, and stars. Both teacher and pupils, together, should note changes in the position and appearance of these objects, make rough measurements, and keep simple records. Then, along with doing these things, they can use books, films, a planetarium, and other teaching aids to *supplement* the real experiences and make them even more meaningful.

SOME IMPORTANT OBJECTIVES

Attitudes and Appreciations to Be Encouraged

A person can gain a feeling of satisfaction and inspiration from observing the sun, moon, and stars.

It is great fun to recognize stars in the sky, such as the North Star and Sirius, and patterns of stars, such as the Big Dipper and Orion.

There is a beautiful orderliness in the motions and changes in the appearance of the sun, moon, and stars.

One need not have a telescope or other expensive equipment, or unusual scientific ability, to learn a good deal about astronomy.

As one continues to observe and learn more about astronomy, the subject becomes increasingly interesting.

Astronomy makes a first-rate hobby, and for those whose interest is especially great, it may become a fascinating life's work.

Most of what one observes about the sun, moon, and stars can also be seen by people all over the earth, and much of it was observed in ancient times.

Things may not actually be as they *appear;* thus, the sun *seems* to be small, the moon to change its shape, all the stars at night to be equally far away, and the earth to stand still while these other objects move in the sky.

Although scientists know much about the universe, there remains a great deal more still to be discovered and understood.

There is no evidence that stars, planets, and other astronomical bodies affect our lives and fortunes in any supernatural way.

Skills and Habits to Be Developed

Avoiding looking directly at the sun, since direct sunlight may do serious damage to one's eyes

Noting the length and direction of sun-caused shadows and their relation to the sun's changing position in the sky

Finding the moon in the daytime, and demonstrating how its apparent shape is related to the direction of the sunlight that reaches it

Recognizing that the moon seems to change its shape in a regular sequence as it moves in relation to the sun

Predicting what the appearance of the moon will be on future dates

Describing how the sun's daily path across the sky changes during the year, and how this change is related to the time and place of sunset and sunrise, the length of daylight, and the seasons

Detecting the motions of the sun, moon, and stars in relation to objects on the earth, and estimating the rates of these apparent motions

Discerning differences in the brightness and color of stars, and in how far they seem to shift in an hour's time

Locating the North Star in the sky, and using it to tell direction

Identifying a few bright stars by name, as well as recognizing some constellations and other star patterns

Distinguishing planets from stars on the basis of recorded observations of their changes in position in relation to the stars

Setting up a true-scale model of the moon and earth, and figuring out how big and how far away a model of the sun would be on the same scale

Placing a globe in sunlight so that it has the same position as the earth in space, and comparing sunlight and shadows on it with those on the earth

Drawing ellipses of various shapes, including some having the same eccentricity as the earth's orbit around the sun

Using correctly, in speaking and writing, such terms as *axis, constellation, diameter, ellipse, focus, gibbous, obtuse, planet, scale model, wane*

Facts and Principles to Be Taught

During each clear day the sun appears to move westward along a path across the sky, and this path gradually changes throughout the year.

The time and place of sunset and sunrise, the length of daylight, and the seasons are all related to the changing position of the sun's daily path across the sky.

The sun's position in the sky at any moment, as well as its daily path across the sky, are different in different places on the earth.

The moon appears to move across the sky in the same general direction as the sun, only a little more slowly.

The apparent shape of the moon depends on how much we can see of its sunlit half, and this is determined by the moon's position in the sky in relation to the sun.

Like the sun and moon, the stars we see at night also seem to move across the sky; all shift in unison and thus keep the same patterns.

The apparent daily motions of the sun, moon, and stars across the sky can be explained by the west-to-east turning of the earth on its axis.

The change in the moon's position in the sky in relation to the sun cannot be due to the turning of the earth; instead, it is caused by the moon's moving around the earth.

Unlike other stars, the North Star changes its position in the sky very little during the night and year; this is interpreted to indicate that it is nearly in line with the earth's axis.

At night, the brighter planets appear much like stars in the sky, but unlike true stars, they keep changing their positions in relation to the stars.

Careful surveying shows the moon to be about one-fourth as big as the earth in diameter, and about 30 earth-diameters away.

Evidence indicates that the sun is a hot ball about 109 times as big as the earth in diameter, and nearly 12,000 earth-diameters away.

Although the earth's orbit around the sun is an ellipse, it is practically a circle; accordingly, seasonal temperature changes on earth cannot be due to changes in the earth's distance from the sun.

SHADOW MOTION

Children should become aware of how the sun appears to move across the sky during the day. Toward this end, one effective way to start is with a contest.

Some sunny morning, take the class outdoors. Set a large can of soil or stones in a level place and stand a tall stick upright in it. Or hammer a stick into the ground. *Ask the pupils not to move the stick, even slightly.* Then mark the tip of its shadow in some way.

Where will the tip of the shadow be after 15 minutes? Let everyone predict the place—*independently*—and mark it with a pebble, button, bottle cap, or bit of tape. *When the time is·up, to whose marker has the tip of the shadow come closest?*

It is fun to repeat this, and also to try it on other days, at different times of the day, and with longer intervals. Meanwhile, pupils can note the movement of other shadows. With practice, they can acquire considerable skill in predicting how shadows shift, and then relate this to the changing position of the sun in the sky.

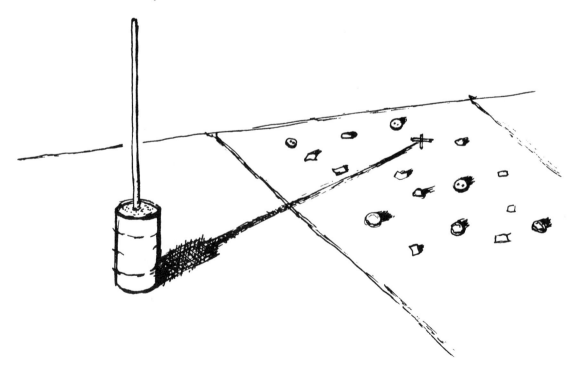

STAR SHIFTS

Nowadays it is often difficult enough to see the stars, let alone observe them move. Nevertheless, every child has a right to have these experiences. Trying to use a planetarium as a substitute for them is a little like trying to have children experience flights of wild geese by showing an animated cartoon!

To provide your class with an opportunity to observe *real* stars shift across the sky, arrange to meet on a clear evening. Choose an open area away from bright lights. Allow for bad weather by setting alternative dates. If school buses cannot be used for this worthwhile purpose, send notes home requesting that each pupil be brought by a parent. Then the parents will share the responsibility for their children—and their children's learning!

Do not be discouraged if only a few come. The ones who do—those who are interested—should not be deprived of a worthwhile experience. And, if the session is kept interesting, the number of sky-watchers may grow at subsequent meetings.

At the meeting ask everyone, both pupils and parents, to select a star of his or her own, and then to walk around until this star appears to touch the top of a tree or pole, or the edge of a roof. The observer should stand upright while sighting the star, and then mark the exact spot where he or she is standing—perhaps with a stone.

Now take 15 minutes to point out to the group some differences in the color and brightness of stars, as well as a few constellations. (See "Star Pictures," page 194.) After this, ask each person to go back to the same spot as before and again sight on his or her star.

Then ask all to share what they have observed. *During the 15 minutes, what did the stars appear to do? In which direction? Was there any overall pattern to the shifts? Did anyone's star show no shift?* (See "North-Star Light," page 193.)

Finally, after another 15 minutes, let the group check whether the stars have continued to shift. *At this rate, how far would they shift in 1 hour? In 6 hours? In 24?*

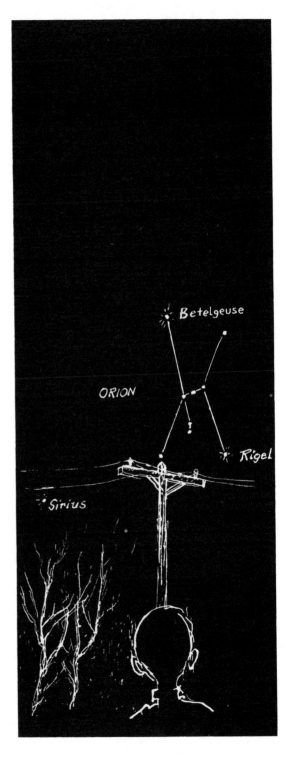

SUNSET PLACE

We often say that the sun rises in the east and sets in the west. Actually, however, this is seldom true—as pupils can discover for themselves.

First locate a readily accessible vantage point with a view toward the west—perhaps at the edge of open land, on a hilltop, or in a tall building. Then draw a sketch of the western skyline as seen from this point, showing trees, buildings, and other landmarks, and duplicate it for the class. Or photograph two color slides of the western skyline, having them overlap slightly so that some distant object directly west of the vantage point appears in both. Project these slides on a long strip of paper taped to the wall and let pupils trace the skyline.

Now ask for two volunteers to go to the vantage point, *accompanied by a parent,* and observe the sun set. They should note the place where it sets and mark this place on the sketch of the skyline. Or, on the tracing, they may show it by taping on a small "sun" of orange paper bearing the date and their names.

Following this, other volunteers—accompanied by parents—should repeat this about once a week. *What change occurs in the place of sunset? Is there also a change in the* time *of sunset?* Pupils should keep a record of this, too.

A month from now, where will the sun set? Two months from now? Let pupils mark the places they predict and then check. In this way they will become aware of a continuous change in the sunset place. Then they can relate this to changes in the seasons, the duration of daylight, and the length of noontime shadows. (See "Shadow Curves.")

When the sun sets south of west, where does it rise? Where does it rise when it sets north of west? Interested volunteers may observe the *sunrise* place, and mark it on a sketch of the *eastern* skyline.

When does the sun rise and set farthest north? Farthest south? When, if ever, does it rise in the east, *and set in the* west?

SHADOW CURVES

After pupils are familiar with the path the sun appears to take across the sky during the day, they should become aware of how this path changes during the year. This may be done as an extension and review of "Shadow Motion" (page 188).

To start, a class should keep a record—for a full sunny day— of the movement of the shadow of a tall stationary object. This object may be a drinking straw standing upright in a lump of modeling clay set on a large sheet of paper or cardboard, weighted down, in a sunny place outdoors. Pupils can take turns marking the tip of its shadow every hour, all day long. Later, they should connect the marks with a smoothly curving line.

Then, to observe how the sun's path across the sky changes during the year, a class should draw such a shadow curve on the first clear day of each month. The pupils should mark the hours and the date on it, and save it. *What difference is there between two shadow curves made during successive months? Why is the curve for December or January, for example, so different from that for October or March?*

Later, as an outgrowth of this experience, older students may swap shadow curves with classes in other schools—ideally in distant places far to the south or north. They may even carry out exchanges with schools in other countries, especially ones located near the equator and on the opposite side of the equator. Besides providing opportunities for learning geography and foreign languages, such an exchange can help the participants in *both* schools better to understand:

The effect of the slant of the sun's rays on the seasons and climate of their home areas, as well as of distant places

The relation of the sun's changing path across the sky to the location of the earth's equator and the tropics of Cancer and Capricorn

The reason for opposite seasons north and south of the equator

DAYTIME MOON

Some children think that the moon can be seen only at night. However, they can discover for themselves that, on *most* clear days, it is visible during daylight, also. This experience is a basic one in astronomy, yet it does not require a telescope or planetarium— or even going out after dark!

First, caution the children never to look directly at the sun. Then let them search for the moon in the morning before school, during outdoor play, and at lunchtime. *Who can be the first to find it?*

When they spot the moon, ask them to compare it with "moons" of assorted shapes cut from paper, and select the one that matches best. After this, let volunteers—perhaps in teams of two— look for the moon each day. Have them take turns for at least a week or two, and keep a record of the moon's changing shape—perhaps on a bulletin board or large calendar.

Each day that someone sees the moon, by day or at night, he or she should add a paper "moon" of matching shape to the record. Or, when the weather is inclement, someone should post a symbol for clouds, rain, or snow. Pupils should record only what they actually *observe,* and leave blanks when necessary. They should not "fill in" just to have the record complete!

From time to time, go outdoors with the class when the moon can be seen by day—preferably when the sun is low in the sky. Let each pupil hold up a tennis ball or other nonglossy ball in sunlight. Help pupils to discern the boundary between the sunlit side of the ball and the part in shadow. Show them how to keep their fingers from getting in the way or casting shadows on the ball. **And again, caution them never to look directly at the sun.**

What fraction of the ball's surface is in sunlight? How much of its sunlit side do you see? Can you notice how this depends on the ball's position in relation to the sun?

The same is true of the moon in the sky. Like the ball, it is in sunlight, but only half of it is sunlit; the other half is in shadow. Usually we can see only the *sunlit* half of the moon or part of it; the half in shadow is generally too dark to be visible. (Actually, it is usually much darker than the shaded side of the ball, which receives a good deal of light by reflection from nearby objects.)

When we see all of the moon's sunlit side, we say that the moon is *full.* When we see half of its sunlit side, we call it a *half moon,* less than half, a *crescent,* and more than half, a *gibbous moon.*

MOON MOTIONS

As a sequel to ''Daytime Moon,'' again take the class outdoors on a clear day when the moon is visible, and have everyone hold up a nonglossy ball in sunlight. **Once more, caution pupils not to look directly at the sun!**

Have each pupil hold up a ball directly in line with the moon in the sky. *How does what you see of the sunlit side of the ball compare in shape with the moon? How can you explain this?*

Now ask everyone to move the ball slowly to the left or right, and note how the appearance of its sunlit side changes. For the same reason, the moon would look different if it were in another part of the sky. *What would it look like if it were farther from the sun in the sky? Nearer to the sun? On the opposite side of the sun?*

Would the straight edge or the round edge of a half moon be nearer to the sun? Would the ''horns'' of a crescent moon point toward *or* away from *the sun? Always?* Pupils can check by moving the ball in relation to the sun.

As observed in ''Daytime Moon,'' the moon seems to change its shape from day to day—to grow larger (*wax*) or to get smaller (*wane*). *As this takes place, how does the moon's position change in relation to the sun? For instance, as the moon waxes, does it get nearer to, or farther from, the sun in the sky? What does the waning moon do?*

A good way to let pupils check this is to have each one extend one arm straight out toward the sun and the other toward the moon. *What kind of angle do your arms make with each other—narrow or wide, acute or obtuse, a right angle or a straight angle?* By recording this angle on successive days, and also the shape the moon seems to have, the class can note how both change. *What causes these changes?*

The sun seems to move across the sky during the day. *Does the moon do this, too? If so, does it follow or lead the sun? Does it always do this, or does it sometimes follow and sometimes lead the sun?* Let pupils find out by observation!

The hour-to-hour shift of the moon across the sky, as of the sun and nighttime stars, is best explained by the turning of the earth. The moon's motion in relation to the sun, however, is not due to the turning of the earth. It is a result of the moon's actually moving around the earth!

NORTH-STAR LIGHT

It is surprising how few children (and adults) are able to find the North Star. Many think that it is very bright; some, that it is the *brightest* star. On the contrary, it is a very ordinary-looking star.

What is outstanding about it is that, while all other stars appear to move across the sky (see "Star Shifts," page 189), the North Star changes its position very little. It stays very close to the same point in the sky all night long (and all day, too) throughout the year. That is why it is so useful for finding direction.

Pupils can easily locate the North Star by using the Big Dipper as a guide. (See "Star Pictures," page 194.) An imaginary line connecting the two end stars (the "Pointers") in the bowl of the Big Dipper, extended five times, almost touches the North Star. Even though the Big Dipper's position changes during the night—and during the year— this relation remains the same.

You can help pupils check on the North Star's position in the sky. Some clear evening stand a stick, nearly as tall as they are, in a large can of sand or stones. Set a screw eye into its upper end. Then place the can and stick so that the North Star, when viewed through the screw eye, appears just above the top of a tall pole. *Caution pupils not to move the stick, even slightly,* and let them check the North Star's position several times during the evening. *Does it shift?*

A screw eye or other peep sight may be permanently fixed in line with the top of a pole and the North Star—perhaps on a post set in the ground by an interested custodian. Then pupils can check the position of the North Star on several evenings during the year. *Does it show any change?*

The beam of North-Star light that comes past the top of the pole and through the peephole is nearly parallel to the earth's axis. (See "Sunlit Earth," page 200.) A line marked on the ground exactly beneath this beam of North-Star light— from the post to the pole—points to true north, or very close to true north. Pupils can compare this direction with *magnetic north,* shown by a compass. (See "Needle Poles," page 149.)

THE **BIG DIPPER**
(Early Evening
in Early Fall)

North
Star

(Early Evening
in Late Winter)

The
Pointers

STAR PICTURES

For thousands of years people have imagined pictures or patterns formed by stars. Finding these is still fun. Children should learn to recognize some of them—in part because they are useful in locating planets in the sky, among the stars. (See "Wanderers in the Sky," page 196.)

To interest your class in star pictures, arrange to meet on a clear evening in a dark, open place. (See "Star Shifts," page 189.) Start by letting the pupils, themselves, imagine pictures or patterns among the stars. They may use a strong flashlight with a narrow beam to point these out. Then, after a while, show them a few *constellations*—star pictures imagined by people of long ago. For this, it helps to have a "star finder" or a chart of the sky in a magazine or book on astronomy.

Some star pictures, such as Orion, are visible during only certain seasons; others, like the Big Dipper, may be visible all year. The constellation Orion, in winter, is very easy to find. The Big Dipper, too, is easy to find—although it is not really a constellation, but only a part of the constellation of the Great Bear, *Ursa Major*.

Each pupil should, at some time or other, line up a star with the top of a pole or tree—and later check its position again from the same spot. (See "Star Shifts.") *What does the star do? Do all the stars in a constellation shift in unison, and thus keep the same pattern? Or does each one shift independently of the others?*

This steady shift of stars across the sky could be because the earth turns around, causing all objects in the sky to *appear* to move in the opposite direction. Or it could be because the stars actually *do* move across the sky, each keeping its place among all the others! *Which is more likely?*

Pupils may also attempt to decide on the order of brightness of the stars in some constellation. On sky charts this order is often indicated by letters of the Greek alphabet. Thus, *Alpha Orionis* is the brightest star in Orion; *Beta Orionis* is next, followed by *Gamma Orionis*.

Finally, pupils may note other differences in the light from stars. *Can you find a star whose light is reddish? Orange or yellow? White or bluish white? Do stars seem to shine with steady light? If not, which ones twinkle more—those low in the sky or those high overhead?*

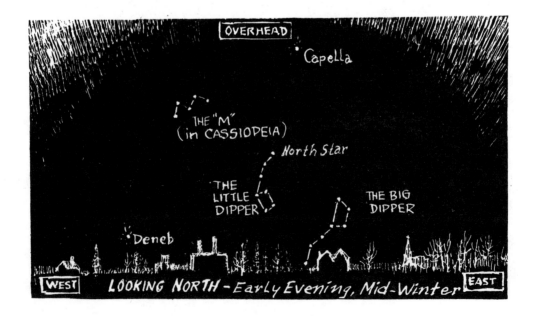

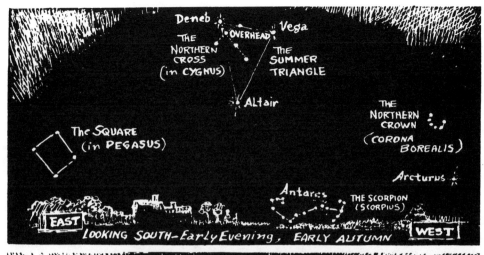

Deneb · OVERHEAD · Vega
THE NORTHERN CROSS (in CYGNUS)
THE SUMMER TRIANGLE
Altair
The SQUARE (in PEGASUS)
THE NORTHERN CROWN (CORONA BOREALIS)
Arcturus
Antares · THE SCORPION (SCORPIUS)
EAST · LOOKING SOUTH—Early Evening, EARLY AUTUMN · WEST

OVERHEAD
THE SEVEN SISTERS (PLEIADES)
THE SQUARE (in PEGASUS)
Aldebaran
· Mira
THE HUNTER (ORION)
Betelgeuse
EAST · LOOKING SOUTH— Early Evening, EARLY WINTER · WEST

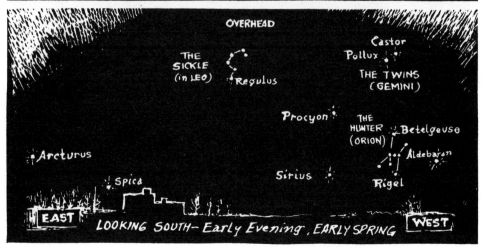

OVERHEAD
Castor
Pollux
THE SICKLE (in LEO)
Regulus
THE TWINS (GEMINI)
Procyon
THE HUNTER (ORION)
Betelgeuse
Arcturus
Aldebaran
Spica
Sirius
Rigel
EAST · LOOKING SOUTH— Early Evening, EARLY SPRING · WEST

WANDERERS IN THE SKY

At night the stars shift across the sky yet keep the same relation to one another. (See "Star Shifts," page 189, and "Star Pictures," page 194.) Other bright objects in the sky, however, slowly change their positions in relation to the stars. Observing these "wander" is a fascinating experience!

The ancient Greeks knew seven "wanderers" in the sky, and our word *planet* comes from their word for "wandering." *What were their seven planets? Of the nine planets known to us, which ones did they* not *know? Why not?*

The ancients had no telescopes, and so they could not see the three faintest planets we know today. Also, to them the earth was *not* a planet! It was not in the sky, and it did not look at all like the other planets. However, they did notice that the moon and sun "wandered" through the constellations. To them, these *were* planets!

Children can readily observe the "wandering" of the moon. They need only make a fairly accurate sketch of it and the stars that appear closest to it; then repeat this the next evening. *How does the moon's position change?*

The sun's "wandering" in relation to the stars we see at night is not directly apparent; when the sun is visible, these stars are not. Still, pupils can note the very first stars to appear after sunset. *Are these the same stars, month after month?*

The "wandering" of Venus and Mars is easy for pupils to observe. They need only find one of these planets* and then mark its position on an accurate sketch of the stars that appear near it. *After a week or so, what change is there in the planet's position in relation to these stars?*

Pupils can note the "wandering" of Jupiter and Saturn, too, in the same way—although these planets move quite slowly in relation to the stars. Mercury, however, never appears far from the sun, and can seldom be seen when stars are visible near it. And Uranus, Neptune, and Pluto are too faint to see without a telescope, although they also "wander" among the stars.

---

*Almanacs and some newspapers and science magazines tell when these planets can be seen. Venus usually appears very bright—in fact, often brighter than any other object in the sky at night. Mars, too, is usually bright, and has a definite reddish color. And both Venus and Mars, like other planets, twinkle less than stars in the same part of the sky.

EARTH-MOON MODEL

Children are excited by the idea of a trip to the moon. Seldom, however, do they have a realistic notion of the distance involved. In large part this is due to the grossly incorrect models and diagrams they so often see in school and elsewhere.

To give them a better concept, help them to set up a simple true-scale model of the earth and moon, accurate as to relative size and distance. Use a globe to represent the earth. Then, for the moon, let pupils find a ball whose diameter is slightly greater than one-fourth the diameter of this globe. This is correct, since the diameter of the moon—almost 2,200 miles—is a little more than one-fourth that of the earth, about 7,900 miles.

Thus, if an 8-inch globe is used for the earth, a ball to represent the moon should be slightly more than 2 inches across; 2¼ inches is close. Or, with a globe 30 centimeters in diameter, an 8-centimeter ball is fine. (Pupils can find the diameter of a ball by squeezing it between two boxes or blocks, and measuring how far these are apart. Slight inaccuracies are not serious.)

Now pupils should put the globe and ball at the ends of a string about 30 times as long as the diameter of the globe. This distance is correct on this scale, since the average distance between the moon and the earth—nearly 240,000 miles—is about 30 times the earth's diameter. Thus, with an 8-inch globe, the string should be about 30 × 8 inches, or 20 feet, in length; with a 30-centimeter globe, about 900 centimeters—9 meters—long.

Could you add a third ball to this earth-moon model, to represent the sun? How big, and how far from the globe, should it be? (The sun is about 865,000 miles across—109 times the earth's diameter. Its distance from the earth averages almost 93,000,000 miles—nearly 12,000 times the earth's diameter!)

On this same scale, how far from the globe should a model be of a man-made satellite that is 100 miles up? About how big should this model be? Which way should it be moving?

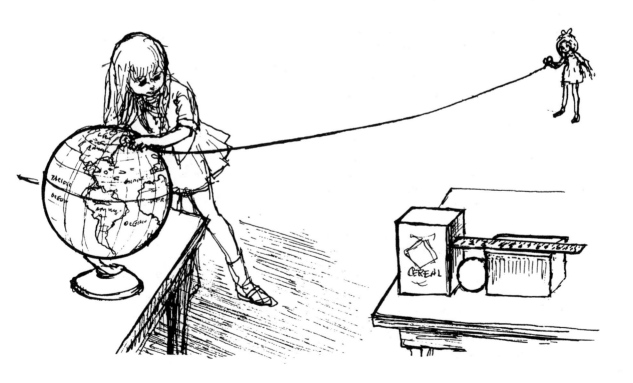

RUN AROUND THE SUN

Many pupils (and teachers) "know" that the path, or *orbit,* of the earth around the sun is an *ellipse.* If they were to sketch it, they usually would draw something like the ellipse below, at *A.*

However, a little thought will show that this sketch is highly inaccurate. If the earth's orbit around the sun *were* of this shape, the distance between the sun and the earth would vary a great deal during the year. Consequently, how big the sun appears to be in the sky—and how much warmth the earth *as a whole* receives from it—would change considerably from month to month. This is not the case!

The fact is that the apparent size of the sun in the sky changes very little during the year. This shows that the distance between the sun and the earth must stay nearly the same. And so, the earth's orbit around the sun—although an ellipse—must be practically a circle. (The notion that the orbit is *highly* elliptical is perpetuated by diagrams that picture it as though viewed obliquely—like the rim of a cup seen from across the table.)

It is contrary to the spirit of science merely to *indoctrinate* pupils with the dictum that the earth travels around the sun once a year, in an elliptical orbit. Instead, they should be helped to understand the *evidence* on which this idea is based. However, if the concept of the earth's orbit is taught at all, it should be presented correctly from the outset. Pupils should, at least, get an *accurate* impression of the shape of the earth's run around the sun.

Toward this end each may draw ellipses of various shapes, and then one that represents the *true* shape of the earth's orbit, as follows:

1 Using a *square knot,* tie a piece of thin string snugly around a 3- by 5-inch file card, kept flat, to make a loop exactly 3 inches long when stretched out straight.

2 Stick a straight pin upright into the middle of a flat sheet of thick corrugated cardboard, or two such sheets.

3 Place the loop around the pin and pull it taut with the point of a ball-point pen.

4 Hold the pen upright and move it around the pin, keeping the loop taut and flat against the cardboard.

Does the distance between the pin and the pen point change? What is the shape of the figure you draw?

5 Now stick a second pin into the cardboard, about 2 inches from the first.

6 Place the loop around *both* pins, keep it taut with the pen point as before, and move the pen around the pins.

This time is the figure you draw as wide as it is long? Is it a circle, or an ellipse?

7 Next, move the second pin to about 2½ inches from the first, and draw another ellipse.

How does the shape of this ellipse compare with that of the first one you drew? Which is more eccentric?

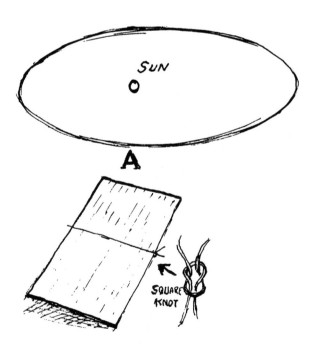

8 Finally, make a whole series of ellipses in this way, leaving the first pin where it is and putting the second at various distances from it.

What happens to the eccentricity *of the ellipses as you make the distance between the pins greater? Smaller? What would the ellipse be like if both pins were so close that they merged into one?*

Now the pupils are ready to draw ellipses that have the same shape as the earth's orbit. For this, have them turn their sheets of cardboard over, and help each one to stick in two pins exactly ³⁄₃₂ inch apart. Next, let each draw an ellipse, using the 3-inch loop. *What does this ellipse look like? How do you* know *that it is an ellipse and not a circle?* Then have each pupil remove one pin and use the loop to draw a true circle around the remaining one, preferably in a different color.

On this scale, the length of the 3-inch loop represents the distance between the sun and the earth, about 93,000,000 miles; thus, ¹⁄₃₂ inch represents roughly 1,000,000 miles. Consequently, on this scale, the sun, about 865,000 miles across, would be the size of a sand grain slightly smaller than ¹⁄₃₂ inch across—roughly the diameter of a pinhole. And the earth would be microscopic!

Like other ellipses, the earth's orbit has two *foci,* represented by the pins. The sun is at one focus; there is no visible object at the other. And so, each pupil may stick a suitable sand grain in one of the pinholes—the one at the center of the circle—and label it "Sun." *According to this scale drawing, where in its orbit does the earth come closest to the sun? During what season is it closest?* (The answer to this last question is: "Winter, in the northern hemisphere—in January"!)

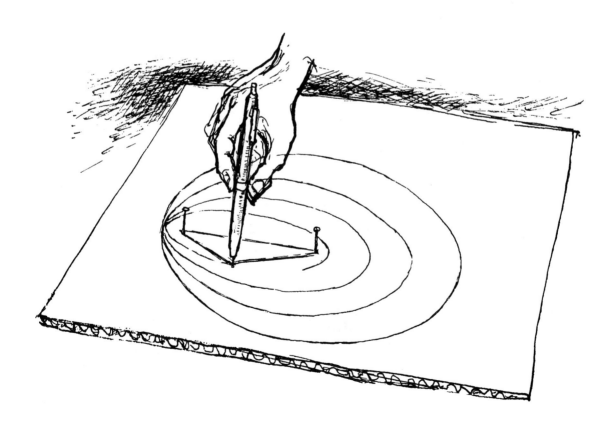

SUNLIT EARTH

What would the earth look like if seen from far out in space? Where on it would there be daytime? Nighttime? Pupils can easily find out.

Take them outdoors on a sunny day and set a globe in sunlight, preferably on blacktop pavement to reduce unwanted reflection. Stand it with its axis in a north-south position. Then rotate it and prop up its base until the place that represents your location on the earth is the *highest* point on the globe. Make sure that the axis is still in a north–south position.

The globe is now in the same position as the real earth in space. Its axis has the same direction as the earth's axis, nearly in line with the North Star. (See "North-Star Light," page 193.) And its sunlit side corresponds to the sunlit side of the earth. *What part of the globe is in sunlight, and what part is in shadow? Therefore, where on the real earth is it day? Where is it night?*

Now stand a toothpick straight up in a small lump of modeling clay at the highest point on the globe—at the place that represents your location on the earth. *How does the direction of the toothpick's shadow compare with the direction of shadows of poles and posts on the ground?*

Stand toothpicks up, also, at a few other places on the globe— each pointing *straight out. Where does such a toothpick cast no shadow? At this place, how do the sun's rays strike the surface of the globe? Where do they strike the earth at this same angle? To someone at that location, where would the sun appear to be in the sky?*

Some sunny day, leave the globe outdoors all day—set up in the same position as the earth. Have the class observe it from time to time. *At a given time, in which countries is the sun rising? Setting? What places have noon? Midnight? Where are children eating breakfast? Supper?*

The globe may be set up like this on occasion, throughout the year. Then pupils can see how sunlight on the earth changes with the seasons. For comparison, they should keep records by taking photographs or making sketches of the globe, from month to month and always at noon standard time.

TOWARD the NORTH STAR

Representative References

Note: Space precludes listing more than a few of the many good books teachers and children will find helpful. A few listed here are out of print (O.P.) but are still to be found in school libraries, and their excellence warrants continued listing. Be sure to consult your school librarian, and publications such as *Science and Children* from National Science Teachers Association and *Appraisal* from the Department of Science and Math Education, Boston University.

COUNTING AND MEASURING

Allison, Linda, *The Reasons for Seasons,* Little, Brown, Boston, 1975. 121 pages. Upper.

Berkeley, Ethel, *Big and Little, Up and Down,* Addison-Wesley, Reading, Mass., 1951. Unpaged. Primary.

Bitter, Gary, and Thomas Metos, *Exploring with Metrics,* Messner, New York, 1975. 62 pages. Intermediate, upper.

Burns, Marilyn, *This Book Is about Time,* Little, Brown, Boston, 1978. 127 pages. Upper, teacher.

Elementary Science Study, *Primary Balancing,* Webster Division, McGraw-Hill, New York, 1976. 112 pages. Teacher.

Leaf, Munro, *Metric Can Be Fun!* Lippincott, New York, 1976. 62 pages. Intermediate, upper.

AIR AND WEATHER

Knight, David, *The First Book of Air,* F. Watts, New York, 1961. 65 pages. Upper.

Moncure, Jane, *What Causes It? A Beginning Book about Weather,* Child's World, Elgin, Ill., 1977. 32 pages. Primary.

Riehl, Herbert, *Introduction to the Atmosphere,* 3d ed., McGraw-Hill, New York, 1978. 340 pages. Teacher.

Sattler, Helen, *Nature's Weather Forecasters,* Nelson, Nashville, Tenn., 1978. 158 pages. Upper, teacher.

Schneider, Herman, *Everyday Weather and How It Works*, McGraw-Hill, New York, 1961. 194 pages. Upper.

Yerian, Cameron, *Projects: Earth and Sky*, Children's Press, Chicago, 1974. 46 pages. Upper.

Zim, Herbert, et al., *Weather*, Golden Press, New York, 1975. 157 pages. Upper, teacher.

PLANTS AND ANIMALS

Comstock, Anna B., *Handbook of Nature Study*, Cornell Press, Ithaca, N.Y., 1939. 909 pages. Teacher.

———, *Ways of the Six-Footed*, Cornell Press, Ithaca, N.Y., 1977. 149 pages. Upper, teacher.

Craig, M. Jean, *Dinosaurs and More Dinosaurs*, Four Winds, New York, 1965. 95 pages. Upper.

Kaufman, Joe, *How . . Born . . Grow . . Work . . Learn*, Golden Press, New York, 1975. 95 pages. Intermediate, upper.

Kessler, Ethel, and Leonard Kessler, *Two, Four, Six, Eight—A Book about Legs*, Dodd, Mead, New York, 1980. Unpaged. Primary.

The Living World, Warwick Press, New York, 1976. 155 pages. Upper, teacher.

Milgrom, Harry, *ABC of Ecology*, Macmillan, New York, 1972. 52 pages. Primary.

Milne, Lorus, and Margery Milne, *Gadabouts and Stick-at-Homes*, Sierra Club, Scribner's, New York, 1980. 108 pages. Intermediate, upper.

Mitchell, John, and Massachusetts Audubon Society, *The Curious Naturalist*, Prentice-Hall, Englewood Cliffs, N.J., 1980. 196+ pages. Upper.

Newton, James, *Forest Log*, Thomas Y. Crowell, New York, 1980. 26 pages. Primary, intermediate.

Palmer, E. L., and H. S. Fowler, *Fieldbook of Natural History*, 2d ed., McGraw-Hill, New York, 1975. 779 pages. Teacher.

Pringle, Laurence, *Animals and Their Niches*, Morrow, New York, 1977. 62 pages. Intermediate, upper.

Ranger Rick's Nature Magazine, National Wildlife Federation, Vienna, Va. Monthly periodical. Intermediate, upper.

Selsam, Millicent, and Jerome Wexler, *Eat the Fruit, Plant the Seed*, Morrow, New York, 1980. 48 pages. Intermediate, upper.

Simon, Seymour, *Pets in a Jar*, Penguin, New York, 1975. 92 pages. Upper, teacher.

WATER AND OTHER LIQUIDS

Arnov, Boris, *Water: Experiments to Understand It*, Lothrop, Lee and Shepard, New York, 1980. 63 pages. Upper.

Cartwright, Sally, *Water Is Wet*, Coward-McCann and Geoghegan, New York, 1973. 26 pages. Primary.

Cherrier, Francois, *Fascinating Experiments in Physics*, Sterling, New York, 1978. 94 pages. Upper.

Educational Development Center, *Colored Solutions*, Webster Division, McGraw-Hill, New York, 1970. 45 pages. Teacher.

Simon, Seymour, *Let's Try It Out . . . Wet and Dry*, McGraw-Hill, New York, 1969. 39 pages. Primary.

Stetten, Mary, *Let's Play Science*, Harper & Row, New York, 1979. 85 pages. Primary, intermediate.

POWDERS AND SOLUTIONS

Educational Development Center, *Mystery Powders*, Webster Division, McGraw-Hill, New York, 1967. 19 pages. Teacher.

Freeman, Mae, and Ira Freeman, *Fun with Chemistry*, Random House, New York, 1967. 52 pages. Upper.

Mullin, Virginia, *Chemistry Experiments for Children*, Dover, New York, 1968. 95 pages. Upper.

Shalit, Nathan, *Cup and Saucer Chemistry*, Grosset & Dunlap, New York, 1972. 93 pages. Upper.

Stein, Sara, *The Science Book*, Workman, New York, 1979. 263 pages. Primary, intermediate, upper, teacher.

Vivian, Charles, *Science Experiments and Amusements for Children*, Dover, New York, 1967. 64 pages. Primary, intermediate, upper.

ROCKS AND THE LAND

Adams, G. F., and Jerome Wykoff, *Landforms*, Golden Press, New York, 1971. 155 pages. Upper, teacher.

Chesterman, Charles, *The Audubon Society Field Guide to North American Rocks and Minerals*, Knopf, New York, 1978. 926 pages. Teacher.

Fenton, C. L., *Life Long Ago*, John Day, New York, 1937. (O.P.) 280 pages. Teacher.

Goldreich, Gloria, and Ester Goldreich, *What Can She Be? A Geologist*, Lothrop, Lee and Shepard, New York, 1976. 48 pages. Intermediate, upper.

Hussey, Lois, and Catherine Pessino, *Collecting Small*

Fossils, Thomas Y. Crowell, New York, 1970. 52 pages. Upper.

Matthews, W. H., *Exploring the World of Fossils,* Children's Press, Chicago, 1964. 153 pages. Teacher.

Rutland, Jonathan, *The Violent Earth,* Warwick Press, New York, 1979. 24 pages. Upper.

Simon, Seymour, *The Rock-Hound's Book,* Viking Press, New York, 1973. 75 pages. Upper, teacher.

Zim, Herbert, and Paul Shaffer, *Rocks and Minerals,* Golden Press, New York, 1957. 160 pages. Upper, teacher.

ENVIRONMENT AND CONSERVATION

Behnke, Frances, *The Changing World of Living Things,* Holt, New York, 1972. 150 pages. Upper, teacher.

Blackwelder, Sheila, *Science for All Seasons,* Prentice-Hall, Englewood Cliffs, N.J., 1980. 254 pages. Teachers of primary grades.

Hungerford, Harold, *Ecology: The Circle of Life,* Children's Press, Chicago, 1971. 92 pages. Upper, teacher.

Pringle, Laurence, *Ecology: Science of Survival,* Macmillan, New York, 1971. 152 pages. Upper, teacher.

Schlichting, Harold, *Ecology: The Study of Environment,* Steck-Vaughn, Austin, Tex. 1971. 48 pages. Upper, teacher.

Schwartz, George, *Food Chains and Ecosystems: Ecology for Young Experimenters,* Doubleday, Garden City, N.Y., 1974. 109 pages. Upper.

Sootin, Harry, *Easy Experiments with Water Pollution,* Scholastic Book Services, New York, 1974. 109 pages. Upper.

FORCES AND MOTIONS

Corbett, Scott, *What Makes a Boat Float?* Little, Brown, Boston, 1970. 42 pages. Intermediate, upper.

Fisher, S. H., *Tabletop Science: Physics Experiments for Everyone,* Natural History Press, Garden City, N.Y., 1972. 123 pages. Upper.

Renner, Al, *How to Build a Better Mousetrap Can & Other Experimental Science Fun,* Dodd, Mead, New York, 1977. 128 pages. Upper.

Ruchlis, Hy, *Orbit,* Harper & Row, New York, 1958. 147 pages. Upper, teacher.

Sharp, Elizabeth, *Simple Machines and How They Work,* Random House, New York, 1959. 81 pages. Upper.

Stone, George, *More Science Projects You Can Do,* Prentice-Hall, Englewood Cliffs, N.J., 1970. 112 pages. Upper.

Zaffo, George, *The Giant Nursery Book of Things That Work,* Doubleday, Garden City, N.Y., 1967. 185 pages. Primary.

VIBRATIONS AND SOUNDS

Cherrier, Francois, *Fascinating Experiments in Physics,* Sterling, New York, 1978. 94 pages. Upper, teacher.

Freeman, Ira, *Sound and Ultrasonics,* Random House, New York, 1968. 131 pages. Upper, teacher.

Mandell, Muriel, *Physics Experiments for Children,* Dover, New York, 1968. 95 pages. Upper, teacher.

Navarra, John, *Our Noisy World,* Doubleday, Garden City, N.Y., 1969. 196 pages. Upper, teacher.

Pine, Tillie, and Joseph Levine, *Sounds All Around,* Whittlesey House, McGraw-Hill, New York, O.P. Intermediate, upper.

Stambler, Irwin, *The World of Sounds,* Norton, New York, 1967. Out of print. Upper.

Stevens, S., Fred Warshofsky, and editors of Time-Life Books, *Sound and Hearing,* Time-Life, New York, 1969. 200 pages. Upper, teacher.

Tannenbaum, Beulah, and Myra Stillman, *Understanding Sound,* McGraw-Hill, 1973. Upper.

MAGNETISM AND ELECTRICITY

Branley, Franklyn, and Eleanor Vaughan, *Mickey's Magnet,* Crowell, New York, 1956. 48 pages. Primary.

Educational Development Center, *Batteries and Bulbs,* Webster Division, McGraw-Hill, New York, 1971. Upper, teacher.

Freeman, Mae, *The Real Magnet Book,* Scholastic Book Services, New York, 1967. 63 pages. Primary.

Reuben, Gabriel, *Electricity Experiments for Children,* Dover, New York, 1968. 95 pages. Upper.

Ruchlis, Hy, *Wonders of Electricity,* Harper & Row, New York, 1965. 218 pages. Upper.

Stone, A. H., and Bertram Siegel, *Turned On: A Look at Electricity,* Prentice-Hall, Englewood Cliffs, N.J., 1970. 64 pages. Intermediate, upper.

LIGHT AND OTHER RADIATIONS

Branley, Franklyn, *Light and Darkness,* Crowell, New York, 1975. 33 pages. Primary.

Educational Development Corporation, *Light and Shadows,* Webster Division, McGraw-Hill, New York, 1976. 32 pages. Teacher's guide.

Epstein, Sam, and Beryl Epstein, *Look in the Mirror,* Holiday House, New York, 1973. 40 pages. Primary.

Freeman, Ira, *Science of Light and Radiation,* Random House, New York, 1968. 125 pages. Upper.

Gardner, Robert, *Shadow Science,* Doubleday, Garden City, N.Y., 1976. 124 pages. Primary.

Ruchlis, Hy, *The Wonder of Light,* Harper & Row, New York, 1960. 154 pages. Upper.

HEAT AND ENERGY

Adler, Irving, *Hot and Cold,* John Day, New York, 1974. 144 pages. Upper.

———, *Energy,* John Day, New York, 1970. 48 pages. Intermediate.

Goldin, Augusta, *Oceans of Energy: Reservoir of Power for the Future,* Harcourt Brace Jovanovich, New York, 1980. 160 pages. Upper, teacher.

Hogben, Lancelot, *The Wonderful World of Energy,* Doubleday, Garden City, N.Y., 1968. (O.P.) Intermediate.

Knight, David, *Harnessing the Sun,* Morrow, New York, 1976. 128 pages. Upper, teacher.

Pine, Tillie, *Energy All Around,* McGraw-Hill, New York, 1975. 48 pages. Upper.

Stone, George, *Science Projects You Can Do,* Prentice-Hall, Englewood Cliffs, N.J., 1963. Unpaged. Upper.

Woodburn, John, *The Whole Earth Energy Crisis,* Putnam, New York, 1973. 189 pages. Upper, teacher.

SUN, MOON, AND STARS

Educational Development Corporation, *Where Is the Moon?* Webster Division, McGraw-Hill, New York, 1968. 56 pages. Teacher's guide. (Accompanied by student booklet, *Where Was the Moon?*)

Gallant, Roy, *Exploring the Planets,* Doubleday, Garden City, N.Y., 1967. 121 pages. Upper, teacher.

Jobb, Jamie, *The Night Sky Book,* Little, Brown, Boston, 1977. 127 pages. Upper, teacher.

Luminous Star Finder and Zodiac Dial, Rand McNally, Chicago, 1970. Not paged. Upper, teacher.

Rey, H. A., *Find the Constellations,* Houghton Mifflin, Boston, 1976. 72 pages. Upper.

Weart, Spencer, *How to Build a Sun,* Coward-McCann and Geoghegan, New York, 1970. 91 pages. Upper, teacher.

Zim, Herbert, and Robert Baker, *Stars,* Golden Press, New York, 1975. 158 pages. Upper, teacher.

Index